Planning in Education

Planning in Education

B. W. VAUGHAN

Cambridge University Press
Cambridge
London · New York · Melbourne

Published by the Syndics of the Cambridge University Press
The Pitt Building, Trumpington Street, Cambridge CB2 1RP
Bentley House, 200 Euston Road, London NW1 2DB
32 East 57th Street, New York, NY 10022, USA
296 Beaconsfield Parade, Middle Park, Melbourne 3206, Australia

© Cambridge University Press 1978

First published 1978

Printed in Great Britain
at the University Press
Cambridge

Library of Congress Cataloging in Publication Data
Vaughan, B   W   1921–
Planning in education.
Bibliography: p.
Includes index.
1. Educational planning—United States. 2. School
management and organization—United States. 3. System
analysis. I. Title.
LA210.V38   375'.001   77–82520
ISBN 0 521 21817 9 hard cover
ISBN 0 521 29285 9 paperback

Contents

Preface

For some years I had been concerned that the traditional method of setting out a mathematics syllabus in the form of pages of typescript did not seem to meet the needs of teachers faced with mixed ability classes. While on a course at the University of Birmingham I came upon the operational research technique of network analysis (also known as critical path analysis, or critical path method, CPA or CPM). I used the technique to set out a mathematics syllabus in draft and later modified it in the light of criticisms from teachers, lecturers and post graduate students. From this I produced a network analysis of infant–primary mathematics which my headmaster, Mr A. Richardson allowed me to use and develop in Parochial School, Trowbridge, Wiltshire. During the process of critical evaluation a number of suggestions were made as to how the network might be used to form the basis of the integrated scheme of work, which is described in this book. I was greatly assisted by the active and constructive support and help given by fellow members of staff. Since this method of setting out a syllabus seemed to have certain advantages over the normal layout I then attempted to see how far other subjects might be treated in the same way and these results are also included. During this development I modified the network analysis technique to make it simpler, easier to understand and more suitable for curriculum planning (chapters 4, 5 and 6).

At the same time I was looking closely at some of the organisational problems faced by today's larger schools. In a small school where all the staff are working closely together any planning can be informal. As the sizes of schools increase the need for formal planning techniques becomes urgent.

The networks included in this book show how the network analysis technique may be used to meet this need. Any school or college may easily adapt these networks to meet their special requirements. Colleges of education might find value, too, in introducing the techniques of network analysis to their students (chapter 8).

I would not advocate the use of any of the enclosed networks as they stand without very critical examination. In the case of subject networks, there may be areas of the curriculum not covered as fully as some schools may wish. In other areas there may be items included which are not essential. What I am offering is, in effect, a series of working examples. I would think it best for groups of teachers and students to try to draw up their own networks, even if they are of only a limited section of the syllabus. The disciplined application of logic and formal planning techniques to consideration of the syllabus is an invaluable exercise.

There has been grave national concern about standards in education recently. One could argue strongly that one of the main problems is that schools are not fully aware what individual pupils are doing in mixed ability classes. The technique of network analysis is admirably suited to ensuring that everyone involved in a syllabus knows exactly what he should be doing, and it is comparatively easy to ensure that all the work in the syllabus is done on time, in sequence and correctly. Surely this is exactly what is required to ensure that standards generally are raised, whether standards of work produced by pupils, or in matters of school organisation?

It would be almost impossible to express my

thanks to all those who have helped with such a long-term project but I would like to thank the following people who gave vital help and encouragement: Professor G. Burroughs, Dr L. Biran and Mr C. V. Platts, Mr C. Buckle and Mr W. Wynne Willson at the University of Birmingham. I also received considerable help from postgraduate students among who were Tom Hume, Alan Wood, Tom Wyant, and the late Ken Gray.

While I was working on the project in school, members of the staff all gave help either positively and actively, or by critically examining what was being done so that it could be progressively improved. Those most closely involved were Mrs S. Grilse, Mrs M. Almond, Mr G. Banks and Mr G. Bridges. Mr C. Rushton of the County Education Service also gave advice and practical help. Help and encouragement was also given by Mr M. Ward, and Mr D. Malvern then at Reading University, Dr R. Sumner of the National Foundation for Educational Research and Mr C. Burgess of Science Research Associates.

Dr B. Twaites and Dr A. T. Rogerson of the School Mathematics Project invited me to serve on secondment for a year as an assistant editor with the School Mathematics Project 7–13. The way in which some of the ideas of modified network analysis were used in the planning stages and in the design of a mathematics curriculum are described in Chapter 7.

Finally I must thank Mrs J. Colby who typed the manuscript promptly and with great accuracy.

Priory Park   B. W. Vaughan
Bradford on Avon
May 1978

1 · Developments in education

There have been a large number of changes in schools in the recent past which have had far-reaching effects upon planning and organisation and have called for changes in approach and technique in the classroom. These changes and their implications will be examined in some detail in this first chapter.

These changes have produced problems in two main areas. The first is school organisation. The increase in the sheer size of schools and the numbers of pupils and staff involved calls for more effective methods of planning and organisation. The operational research technique of network analysis has for years been used in industry to solve the many problems associated with complex projects involving men and materials. The technique is described in chapter 2 and applied in various school planning situations in chapter 8.

The second main problem area is curriculum planning; how to set out a curriculum clearly so that all those who use it know not only what to do but also the order in which to teach each item. Following on from that, the problem is how to provide an adequate and effective record-keeping system and include the tests necessary to monitor not only the progress of the individual pupil, but also the overall scheme of work. In chapter 3 a modification of the operational research technique is described which has been found effective in helping to set out a curriculum and as the basis of an integrated scheme.

Chapter 4 describes in detail how such a network analysis of primary mathematics was designed and the scheme implemented. Chapters 5 and 6 set out the way in which curricula for science, English,

history, geography, art and craft could be designed and the way in which the design may also be developed.

The general changes which have affected schools are: 1 Changes in the expectations of society. 2 New theories of learning and teaching. 3 Changes in curriculum content. 4 Changes in the size of schools. 5 Variations in the age of transfer from school to school. 6 Greater mobility of pupils and staff. 7 Raising of the school leaving age. 8 New methods of class organisation. 9 Changes in teaching materials. 10 The introduction of audio-visual techniques. 11 Open-plan schools.

1 Changes in the expectations of society

The time has now long gone when all that was required of pupils by the time they left school was that they could read, write neatly, and carry out quick arithmetical computations. The curriculum has been so radically altered that it can hardly be recognised. There is an emphasis upon the understanding of principles which can be applied, rather than upon techniques which can be used.

So many revolutionary changes have taken place outside schools with the introduction of new technologies and the certainty that there are more to come, that schools have had to change their approaches. Since it cannot be known what tasks pupils will be required to carry out in the future it is essential that they are taught basic principles which can be used whatever circumstances arise.

Reading is no longer a matter of reading aloud clearly, but of learning a variety of techniques which can be applied as circumstances demand.

Mathematics is no longer the production of answers to computation questions; it is concerned with problem solving.

Traditional class lessons are often replaced by setting pupils individual tasks, to find out for themselves solutions to problems which may not have obvious solutions. Projects aimed at gathering data and then analysing it in a systematic manner are also widely used.

2 New theories of learning and teaching

A great deal of thought has been directed to the ways in which children learn. As a result there is fairly general agreement that children learn in a series of steps or stages. These are sometimes referred to as hierarchies of learning. These stages have been roughly identified and have been given different names by different experimenters so that at times it becomes difficult to relate one to another. If the theory is accepted that children learn by handling concrete materials in the early stages of learning and proceed by gradual steps to abstract, higher order learning, then teaching should be geared to this and teaching materials designed to facilitate learning in such a series of stages. If learning is geared to the stages which have been identified and the teaching materials present each pupil with the experiences which will facilitate individual learning, then mastery and understanding should be speeded up.

These ideas have had a wide influence upon teaching methods, techniques and teaching materials. If we are to think about each pupil proceeding through these series of hierarchies of learning at a pace best suited to his individual needs, then class teaching is no longer effective. Some form of individual teaching is required. In the same way a standard text book no longer meets the needs of the individual pupil. Even if text books are graded for children of different levels of ability they will not suffice. It is a complex matter to arrange for each pupil to have the variety of experiences which best meets his needs, at the correct time, and including sufficient practice in basic skills and techniques. Such a flexible course cannot be found in bound books. For this reason schools have tried to meet the needs of the pupils by arranging courses which are more adaptable. The project approach has been tried out with varying success. There are real difficulties in arranging this sort of teaching and keeping a check of the exact work done by each pupil. It is easy enough to start an open-ended project, but how is the teacher to ensure that each pupil gains the experiences necessary for his progress, how to monitor, how to record, and what to guide the pupil to tackle next? All this has to be sorted out if the project approach is not to dissolve into a diffuse and inefficient method producing minimum results.

Current theories of learning and teaching apart, it is still necessary to decide a starting point, and then decide what shall follow. At some stage decisions have to be taken about the order in which the curriculum is to be presented and the sequence of items in the syllabus.

3 Changes in curriculum content

In many schools for older pupils most teachers divided the subject syllabus into a series of lessons based on a standard text book. The pupils listened, did exercises to reinforce the subject matter of the lesson, and then did further extension work and practice as homework. Now very often the formal teaching techniques which are suitable for able pupils have been adapted, and open-ended exploratory techniques have been used effectively. There

have also been attempts to adopt a more individualised approach so that the teaching more effectively meets the needs of the individual pupil. This emphasis upon what the pupil learns, rather than what the teacher tries to teach, will be dealt with at length in a later chapter.

In schools for younger pupils the syllabus used to be very restricted. It is now much wider ranging and includes much that was once to be found in the curriculum for older pupils. Teaching methods too have changed. Pupils now are engaged in far more practical work, and work on their own. They frequently work as individuals or in small groups. A far wider range of apparatus and materials is deployed.

British primary schools have had to contend with such changes as the abandonment of the farthing (¼d) and then the introduction of decimal currency in 1971. Most countries used the old imperial system of weights and measures and are now adopting SI units. This is not yet complete but will be by the time children now in school have left.

Changes like these have made some earlier text books out-of-date. The replacement which ought to have taken place has not been possible because shortage of funds will not usually allow the wholesale discarding and replacement of sets of books throughout a school. Therefore there has been a policy of gradual replacement of out-of-date books.

4 Changes in the size of schools

Older pupils have been affected by the change which has taken place in the sheer size of schools. It has been thought that to make it possible to have a sufficient variety of courses at advanced level, schools for older children should have over a thousand pupils. The size of such schools has however posed other problems which may not have been foreseen. There comes a time when the benefits which obtain with an increase in size are outweighed by other considerations. There are large

social gains in having a school of a size where staff and pupils know each other. The fact that size in theory allows for a wide variety of courses to be arranged for pupils, is often offset by the timetabling difficulty involved. This is often of such proportions that computers are being increasingly used for timetabling. Even so pupils can find that their courses are arranged not by their desires, but by timetabling constraints.

5 Variations in the age of transfer

The age of transfer from school to school is now so varied that it is impossible to make any sort of generalisation as to what is likely to take place in the future. It is quite clear that the more there is variety of age of transfer, the greater importance must be attached to some method of recording the achievements of pupils, to enable them to be correctly placed in courses suited to their needs after transfer.

In many cases schools have changed from being single sex to having boys and girls in the same buildings. This too has meant a change in attitude and course structure.

6 Greater mobility of pupils and staff

At a time when there is a tendency for larger industrial units to be formed from the amalgamation of smaller firms there is an increased mobility of parents. Very often promotion entails a transfer from one town or city to another. This means that a significant number of pupils are moved at least once in their school career and the possible permutations of schools they may attend is very wide indeed. This too is an argument for having an agreed common curriculum for the basic subjects, linked to an adequate record-keeping system.

At the same time that pupils have become more mobile, so too have teachers. The salary scales are so arranged that in order to gain varied experience and promotion teachers have every inducement to move from one school to another, especially in city centres where constant changes of staff are a severe handicap to the efficient running of schools. While one would not wish to handicap teachers, and to slow down their chances of promotion, there is value in the continuation of a teacher's service in a school. In any case if the school is to continue to operate with efficiency, much careful planning must go into the courses arranged for pupils so that possible changes of staff have the minimum of effect. This can only be done if the courses are planned with great care, are clearly laid out, and if record-keeping is adequate and effective.

If records show clearly what each pupil has done and with what success he has tackled each item in the syllabus, then any new teacher can take over the supervision of the work done by the pupil with minimum disruption.

7 Raising of the school leaving age

The raising of the school leaving age calls for new approaches. Courses designed for the more academic are quite unsuitable for the less able pupil staying longer at school. A variety of courses has had to be designed at different levels. In Britain new types of examination have been designed to test the success of the new courses, and to give the pupils some measure of their attainment.

8 New methods of class organisation

Some schools have classes of thirty or so pupils of the same age but different abilities under the care of one teacher. Others have a group of about sixty pupils under the care of two teachers acting as a team, or about ninety pupils in an area under the care of three teachers again working as a team. They may be in separate classrooms, in two rooms adjoining which can be thrown open, or working in a large area in an open-plan school. The arrangement of the pupils may be in years, or by vertical streaming in which, for example, a class may con-

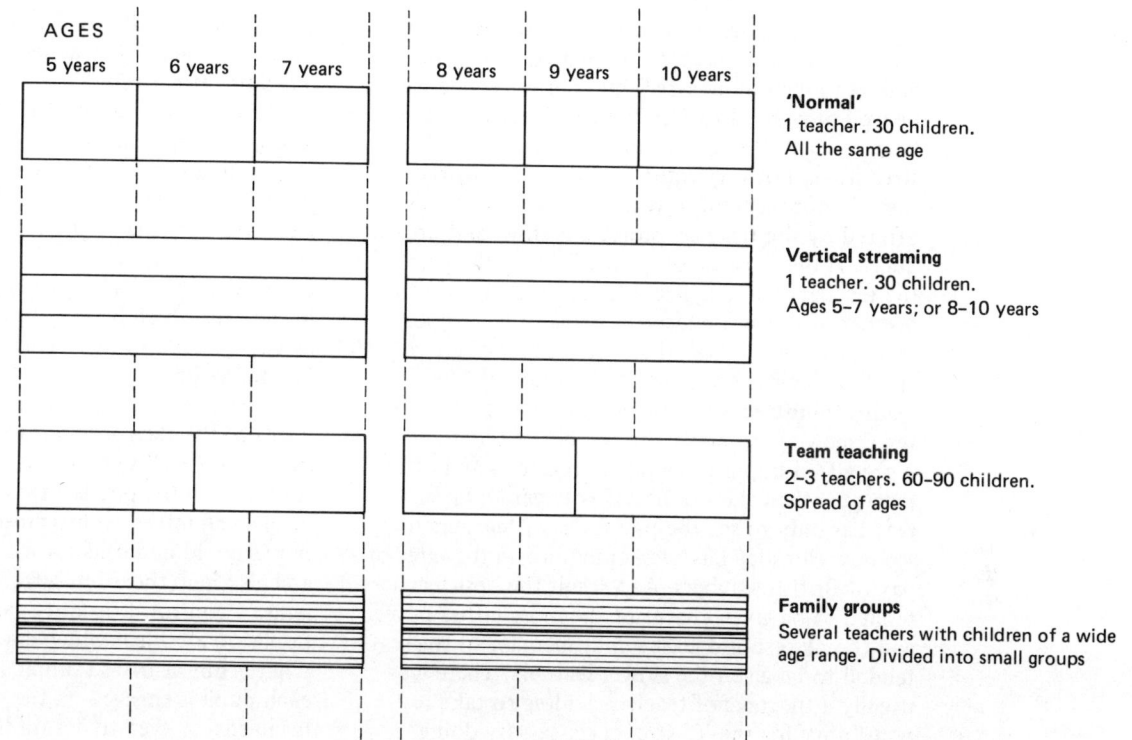

Fig. 1. *A diagram of some of the methods of internal organisation of schools for younger children. A school may use more than one of these systems.*

tain pupils from age five to seven, or seven to nine. Some schools arrange the pupils in 'family groups' with a mixture of older and younger pupils in small groups. There are many variations. Whatever the reasons for change, stated and unstated, today the majority of schools in Britain for younger children are unstreamed, and many schools for older pupils as well.

In vertical streaming and family grouping instead of dividing the school into years of pupils of the same age, and then into classes of pupils of roughly equal size, the school is divided into classes which pay no regard to age or ability.

It used to be the standard practice in all schools of any size to divide the children into streams in accordance with their ability. It is now widely argued that streaming tends to maintain and even increase the divisions between the gifted and the less able. It is also felt that if children are not streamed their social behaviour is often improved. Thus a desire to minimise the differences between children, and an equally strong desire to improve behaviour made many head teachers decide that in the interests of the majority of pupils, it would be better not to stream their classes at all.

There may well have been other motives than social

justice behind some of these changes. There is no doubt that most teachers find teaching brighter children much more satisfying. Certainly one has to be well motivated to face the comparative disincentives to teach the less gifted. Pupils in low ability streams are not very interested in school work, they do not respond so readily to the incentives offered by the teacher, progress is slow, and produced at the expense of great effort on the part of the teacher. There is also a tendency to regard the teacher of slower children as being of less importance, and less skilled, than the teacher of gifted children. This is not an attitude which will be readily admitted, but it is present nevertheless. In the same way generally the teacher of older and more gifted pupils has a higher regard — and often reward — than the teacher of younger children. One has only to see the pay scales of teachers to see how this idea has been embodied in the rates of pay of British teachers. As a result the best teachers tended to be given charge of the more gifted pupils, and the newest and least senior member of the staff tended to be given the slower learners. There was usually a shortage of teachers willing to take responsibility for the 'C' stream classes. By doing away with streaming a head teacher could solve the problem which had to be regularly faced, of finding a teacher who was willing to take over the less gifted classes.

The majority of schools for younger pupils are now unstreamed, though the organisation of the schools varies considerably.

It is clear that such changes create new problems which may not have been foreseen. These problems need close examination.

If a teacher is faced by a class all the same age, and with an ability range which has been minimised by streaming, the class lesson can be the mode. The teacher can explain any new materials in a way which is geared to the needs of the main part of the class, and after examples and explanations most of the pupils will be able to understand. Most text books are written with this in mind, and most

schools have a syllabus designed to this end. In the course of a year the teacher would cover the syllabus quite effectively.

However, with an unstreamed class, the class lesson no longer suffices. The ability range in a class will now be very wide indeed. Thus a lesson geared to the needs of the middle group of the class will bore the brighter pupils, and be beyond the grasp of the less able. Even so, many teachers, faced with this challenging situation, dealt with it in the way which had been effective with streamed classes, and continued to deliver lessons as they had for years.

Others in an attempt to meet the widely varied needs of the class adopted an open-ended approach of the 'project' type. In this a class project was started with the pupils either working individually or in small mixed ability groups. Thus the brighter pupils could undertake work at their own level, and also help those less gifted. Here one faces problems of control. It is fairly easy to start a class of pupils on a project which is worthy of their efforts. What is not nearly so simple is to make certain that each pupil is engaged on the work best suited to his needs, or even to ensure that those who are lazy are engaged in much work of any sort at all. There may be great value in projects of various kinds, but it is very difficult to measure the results obtained, if it is possible at all. The work done by every pupil needs to be monitored if project work is to operate successfully. It may be a valuable and worthwhile exercise, or it may be a waste of time for a large number of pupils, but the teacher must know what methods are being effective in their classroom. Project work if it is to be used should be very carefully planned and supervised.

Many teachers have realised the implications of the unstreamed class and have tried to devise ways in which their pupils can do individual work designed to meet the very varied needs. At once acute problems arise. It is comparatively easy to start a class on work which is suited to pupils' individual needs. Very soon however the quicker pupils, or

those given a fairly simple activity, have completed their assignment; this needs checking, and their next assignment needs to be set. What started as a trickle of pupils wanting their work checked and their next task assigned, soon becomes a flood. A queue forms, and the teacher is under pressure to keep pace. He is faced with demands not only to check the completion and correctness of work done, but an instant decision as to the next work best suited to the needs of each pupil. In the midst of this, questions will be continually asked when pupils want a piece of equipment, new materials, or an explanation of points not understood. Many teachers who start off convinced of the need for individual teaching in their unstreamed class soon become pressurised. The usual syllabus of a number of typed pages is not really much help in this situation. When it is necessary to start a pupil on his next piece of work, one just does not have the time to turn over pages of typescript, or even a book, no matter how well indexed.

The more freedom the pupil is given to pursue his individual course, the greater the attention which must be given to planning the course in advance. What is needed is a mainstream course, with upper and lower levels for the more able and the slower pupils, and some method of checking each activity so that a pupil can be guided to the level he needs to undertake at that particular stage of his school life. One cannot assume that, because a pupil is bright, he will be equally bright at every aspect of a subject, nor that he will always work at the same pace. Children vary in their speed of development, and this must be taken into account in any planning.

A result of the variety of organisation in schools for younger pupils is that schools for older pupils are faced with an intake which is even more varied in attainment than before. Some senior schools try to take account of this by having a number of streams or sets. In others they have adopted an individualised approach. They have recognised the variety of ability and attainment of their intake and have tried to arrange for each pupil to carry on from where he left off as far as this can be arranged.

There should be close consultation between schools to ensure that courses of study are knit one into another if pupils are to make the change from one school to another smoothly. There ought also to be standard record-keeping systems which will enable pupils to proceed with the minimum disruption. There was a time when courses could be geared to the needs of either primary or secondary pupils. Now that the age of transfer is so varied there is a need for courses to be planned to cover a wider range so that they are common to both primary and secondary pupils. One example of effective planning in this area is the School Mathematics Project which, with the publication of SMP 7–13, will be common to both primary and secondary pupils.

9 Changes in teaching materials

There is another requirement to be met before any individualised scheme can operate effectively. There must be effective teaching material. Once the teacher decides that the pupils are to undertake courses best suited to their individual needs, the teacher can no longer be responsible for teaching the contents of the syllabus to each individual pupil as this would be physically impossible. There is however a good case to be made for class lessons to cover certain key points in the syllabus, and these could perhaps be best taught by a teacher teaching the class as a whole. It is highly likely that the best method to employ will prove to be a combination of class lessons, during which the teacher takes the class as a whole, or at least a large part of the class, combined with individual working, in a proportion which experience or experiment will decide. Class lessons enable teachers to establish rapport with pupils, and there are times when a two-way flow of ideas and questions has great value. There is too a value in class lessons to create a group unity, or a community spirit.

Since the teacher cannot possibly teach the major part of the syllabus once individual work is the mode in an unstreamed class, teaching material must be provided. The exact nature of this teaching material must be a decision to be taken by head and class teachers together. At present there is very little on the market, and much more thought ought to be given to the provision of such material. If it is to be provided teachers will have to be convinced that the material will teach what it purports to teach. No teacher should abdicate his rôle as teacher to teaching material, unless he is first assured that there will be no loss to the pupil in his care.

10 The introduction of audio-visual techniques

At one time most schools used broadcast radio programmes, some had ciné and slide projectors, and that was all, except in a few rare cases. Today a very wide range of audio-visual aids is not only available, but in constant use in schools. Each new piece of equipment which is introduced has some undoubted value as an educational aid. The teacher has first to learn how to operate the equipment, then to integrate the use of the equipment into the work of his particular class. In many cases it has proved easier to design and market equipment, than to design adequate and effective teaching material to use with it. Thus if a teacher has been convinced of the value of a particular piece of equipment, he has had to design and produce his own teaching material to use with it.

Many audio-visual aids call for special conditions, such as a blacked-out room or a room free from extraneous noise. This is a real problem in many schools. Very often only one or two rooms have adequate black-out, and so the use of audio-visual aids requires a timetable for the use of that room or rooms.

There is, too, another good reason for more efficient timetabling, as many audio-visual aids can be used with larger groups, and more than one class can be combined when the material to be presented is relevant to their needs.

11 Open-plan schools

Once the system in schools became divorced from the idea of groups in age classes, with year groups divided into streams according to ability, there were a number of interesting changes in the architecture of schools and the organisation within them. It is now quite common to have schools built 'open-plan'. There are no walls between classrooms, and so the school consists of a small number of open areas. As a result the teachers work in small teams and there may be over one hundred pupils in an area, supervised by three or four teachers, often with ancillary help.

In this sort of school, full use can be made of the particular skills of individual teachers. Pupils can go to the area in which the specialist teacher in art and craft can supervise their work, and the necessary materials and equipment can be grouped in this area. The teacher who is skilled in the use of audio-visual equipment can operate in another area, and so on. In this way the best and most economical use can be made of the skills of the staff, and the apparatus and equipment can be deployed most effectively.

There is no doubt that there are certain advantages in the open-plan school in which team teaching is employed. However, all that has been said about the problems of individual courses for pupils applies even more in such schools. Much careful thought must be given to planning the courses, to effective methods of monitoring progress, and to ensuring that the pupils are actually doing the work designed for them.

It is clear that the problems outlined above call for better planning and organisation both in the development and teaching of curricula and in school administration. In order to monitor the progress of each pupil, some sort of simple record-keeping system is necessary. This will not only enable the teacher to ensure that the pupil is following the course best suited to his needs, but also enable him to keep a constant check on the progress the pupil is making. It may also act as a guide to the next relevant activity.

There is also a need for records in order to keep a check on the effectiveness of the syllabus, the teaching which is based upon it, and the teaching materials used. A continual process of improvement and modification should be based upon the records. There should also be a series of tests to ensure that what is intended to be taught is, in point of fact, being taught.

If the record system is to be used at all by a teacher who will be hard pressed to keep up with the continual demands on his time, it must be simple, and require the minimum of routine clerical work.

In any form of individualised teaching scheme it is necessary to teach pupils to be self-reliant. They must be taught how to find the teaching materials they need, and then to find and sensibly use a wide variety of apparatus and equipment. What is perhaps even more important is to have some systematic way of checking the return of teaching materials, apparatus and equipment. If materials are to be returned there must be a system of storage and retrieval which is simple but effective.

Those who are mainly interested in school organisation should turn to chapter 2 where the technique of network analysis is explained and to chapter 8 in which the technique is demonstrated as an aid to planning and organisation.

Those who are concerned with curriculum design and the implementation of schemes of work should follow chapter 2 with chapter 3 which explains how a simplification of network analysis technique may be used to set out a curriculum. Chapter 4 shows in detail how to base a syllabus on such a network analysis of the curriculum, taking primary mathematics as an example. Chap-

ters 5 and 6 show how the same method may be used to implement similar schemes of work in other subjects. Chapter 7 shows how the planning techniques described were applied to the organisation, design and development of the School Mathematics Project's 7–13 course.

The industrial technique of network analysis is explained in chapter 2 in more detail than is usually required for educational planning but it is hoped that this account will be found to be of value in its own right as well as for the insights it gives into the development and application of the simplified techniques with which most of the book is concerned.

2 · Network analysis

The operational research technique of network analysis was developed in order to control and organise complex operations in industry. The technique is sometimes called PERT (programmed evaluation and review technique), critical path analysis, or critical path method.

The planning and organisation of projects which involve less than about fifty activities can normally be carried in the head and so one would not use a technique like network analysis. However, in order to explain the technique an example has been chosen which is likely to be familiar to most teachers in school. There can be few teachers who have not shown a film, film strip or slides. Most teachers will know only too well that although the showing of visual material is fairly simple, any slip-up in planning can result in chaos, for pupils will not just sit quietly while a new lamp is found or a film re-threaded. Having shown how the technique applies to a project involving comparatively few activities it will be easy to transfer the method to more complex operations in school. Teachers will know, of course, that there is one aspect of planning which has to be taken into account when organising school activities, and which does not affect industrial projects. This is the fact that pupils do not behave like miniature adults, and so plans should be made to ensure that at all times they are adequately supervised.

Let us look at the network analysis technique step by step, applying it to the showing of a film:

1 List activities

List all the activities involved in the project and arrange them in logical order. Although not all of the following take place every time a film is shown, it is worth listing most possible activities.

(a) Order the film, or obtain it from store
(b) Arrange chairs for the audience
(c) Erect screen
(d) Erect projector
(e) Adjust screen
(f) Check that a spare lamp is available (There is nothing worse than starting to show a film and then having the lamp fail.)
(g) Insert film and check that it is working freely. Check alignment and focussing
(h) Seat audience
(i) Show film
(j) Store chairs
(k) Store screen
(l) Store projector

2 The arrow diagram

Draw an arrow diagram to show the order in which the activities should take place and check for inconsistencies and to ensure that the activities are in the correct order. There are certain conventions which are used in drawing the arrow diagram which forms the network and these are as follows:

1 Each activity is shown by means of an arrow (see figure 2).
   The arrow always points from left to right. The length of the arrow is no indication of the length of the activity.
2 Each arrow has a circle, sometimes called a 'node' or an 'event' at the start and finish.

The points of the arrow are usually omitted as they are not needed.
The nodes are numbered for identification. The number at the start of an activity is always smaller than that at the end of the activity and each activity has unique numbers.
The numbers need not be in strict sequence. In fact it is wise to leave out some numbers at intervals to make alterations and additions easier (see figure 3).
'Order film' is now activity number: 1–3.
3 All activities shown by arrows entering a node must be completed before those shown as starting after the node, commence (see figure 3a).

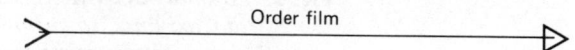

Fig. 2. *An activity arrow.*

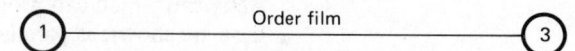

Fig. 3. *Activity arrow with nodes, each enclosing a reference number.*

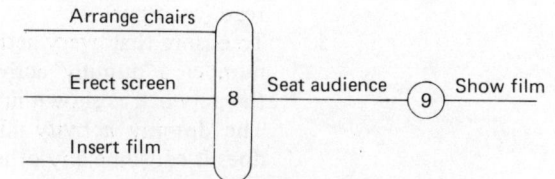

Fig. 3a. *All activities leading to node 8 must be completed before activity 8–9 starts.*

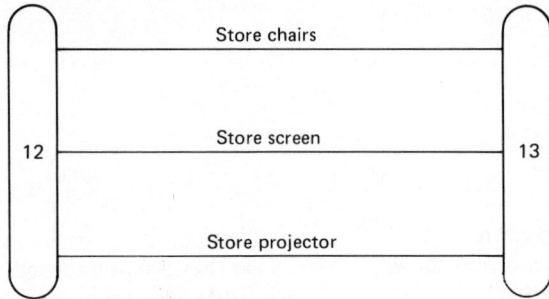

Fig. 4. *Parallel activity arrows to ensure legibility.*

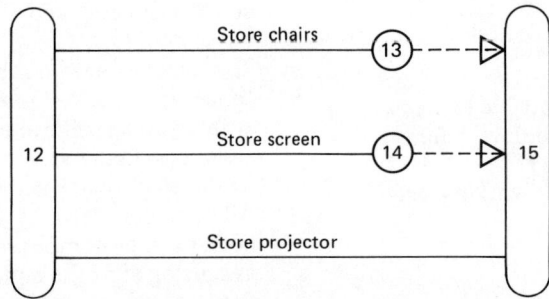

Fig. 5. *Dummy activities inserted, shown as dotted lines, to ensure each activity has a unique number.*

4   Activities which can proceed at the same time are shown as parallel arrows.
When activities are carried out in parallel, elongated nodes are used instead of circles. This is a convenience in drawing and ensures that the name of the activity can be easily read (see figure 4).

5   To ensure that every activity has a unique number, a 'dummy' activity is inserted into the network as shown in figure 5.
The 'dummy' activity takes up no time nor does it consume any other resources. It is merely a convention to enable all activities to have their own unique number. The dummy activities are shown as dotted arrows

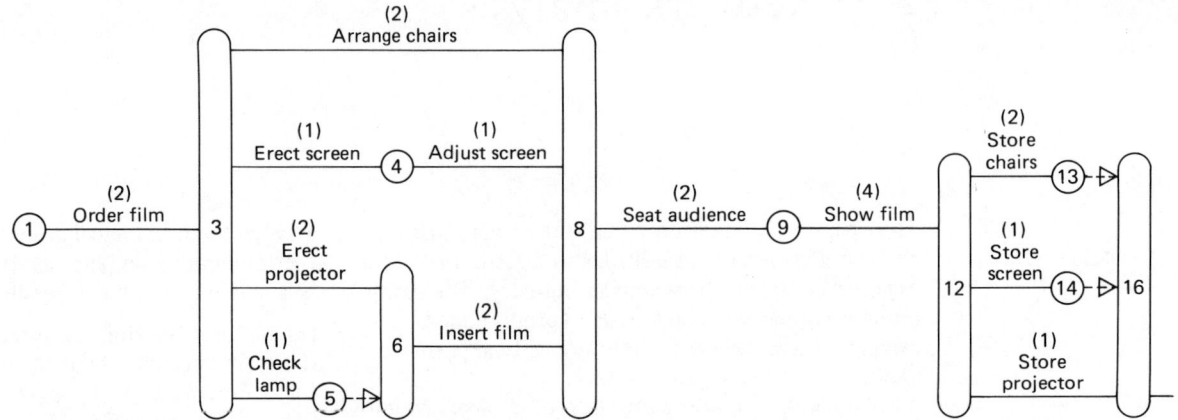

Fig. 6. *Completed network of showing a film to illustrate network drawing conventions.*

(13–15 and 14–15) and the activities are now numbered individually 12–13, 12–14 and 12–15.
The whole network should then resemble the diagram in figure 6.

3   Allocation of resources

One of the conventions of network drawing is that it is assumed that there are no limits to the resources available. It is assumed that there will be no time limits and that equipment and manpower are also unlimited. It is only after the design of the network that resources are considered.

1   Time

One of the most crucial elements of any planning is the time taken to complete a project. It can be roughly estimated by adding together the time taken for each individual activity, but this does not give a true picture as it takes no account of parallel activities. The drawing of a network identifies these parallel activities. A more accurate total can be calculated and the 'critical path' can be found. This is discussed in detail below.

2   Manpower

Once the time taken by activities has been considered and the critical path identified, the allocation of other resources can be decided. Chief among these is the deployment of manpower. If necessary the number of persons employed may be increased either for a time, or for the whole project. The exact place where extra persons may be best used is of prime importance.

3   Equipment

The other main resource which needs consideration is equipment. In the present financial climate when, more than ever, the most effective and economical use of resources in schools is required, the type of planning and organisation involved in network analysis can be a very valuable aid to economy.

Allocation of resources — time

Once the arrow diagram is drawn and has been thoroughly checked the allocation of resources can take place.

The first task is to estimate the time each activity will take. The list might look like figure 7.

	Activity	Minutes of time consumed	Activity numbers
1.	Order film	10	1–3
2.	Arrange chairs	10	3–8
3.	Erect screen	5	3–4
4.	Erect projector	10	3–6
5.	Adjust screen	5	4–8
6.	Check spare lamp	5	3–5
7.	Insert and check film	10	6–8
8.	Seat audience	10	8–9
9.	Show film	20	9–12
10.	Store chairs	10	12–13
11.	Store screen	5	12–14
12.	Store projector	5	12–15

Fig. 7. *Activity list showing the amount of time taken by each activity.*

For a simple series of activities like this the time taken can be estimated fairly well. For more complex series there will be a wider margin of error depending upon variables which are not easy to calculate. Although this sort of consideration is not needed for the film-showing network, an explanation of the method of estimating the time taken by activities will be needed so that it can be used in more complex projects.

Time estimation. When an activity takes place on a number of separate occasions and is carried out by different people and in different circumstances, the time taken depends upon such factors as the skill of the operator, availability of apparatus and materials, and so on.

It has been found that the time taken by repeated activities in industry follows a beta distribution, that is that the most likely trend will be towards an optimistic rather than a pessimistic estimate (W. Clark, *The Gantt Chart*, Pitman, 1952). A full discussion of the various statistical methods and their strengths and drawbacks can be found in T. Cass, *Statistical Methods in Management* and J. E. Kelly and M. R. Walker, *Proceedings of Eastern Joint Computer Conference 1959*, USA.

The formula used here is not particularly sensitive and might not suffice for critical estimates used in industry but will suffice for planning in schools where any minutes saved might well be offset by staff time taken in more complicated computations.

These variations can be taken into account by using the following formula:

$$\frac{B + 4C + A}{6} = \text{time estimate.}$$

A Is the most optimistic time assuming that all conditions are favourable, and the operator is skilled.

B The most pessimistic estimate and assuming that all conditions are unfavourable, and an unskilled operator.

C The most likely time. This would be with a fairly skilled operator, and with some hindrances, but without any really serious hold ups.

In our example a reasonable estimate for inserting film by a skilled operator might be 5 minutes (i.e. $A = 5$).

For an unskilled operator who had to read the instructions, and possibly tangles the film and has to reinsert it, the time might be 13 minutes (i.e. $B = 13$).

The most likely time might be 7 minutes (i.e. $C = 7$.

$$\frac{B + 4C + A}{6} = \frac{13 + (4 \times 7) + 7}{6} = \frac{48}{6} = 8 \text{ minutes.}$$

When the time taken to complete each individual activity has been calculated they are entered in a table, like that in figure 7.

The time taken by activities in school does not need more sophisticated methods of time estimation since most of the activities are repeated year after year, and one has a wide range of experience upon which to draw. The above calculation involves the use of the time unit of one minute. Whatever unit is most convenient may be used so long as the same one is used throughout the network.

The critical path

Once the time taken by each separate activity has been estimated the total time taken by the whole project can be easily worked out by progressive addition. The progressive addition of the activity times is entered in the left-hand side of each oval 'time box'. The estimated times are given, bracketed, in five minute units (see figure 8).

The first entry will be to insert a figure 0 in the left-hand side of the oval immediately below node 1 (entry 1). The time taken by activity 1–3, ordering the film, is added to this 0 and the total inserted in the oval at the foot of elongated node number 3 (entry 2).

The next entries are as follows (see figure 9).

To the total entered in the oval at the foot of node 3:

Entry A add the time taken by activity 3–4 (1 unit) and enter the total, 3 units, in the left-hand side of the oval above node 4;

Entry B add the time taken by activity 3–6, of

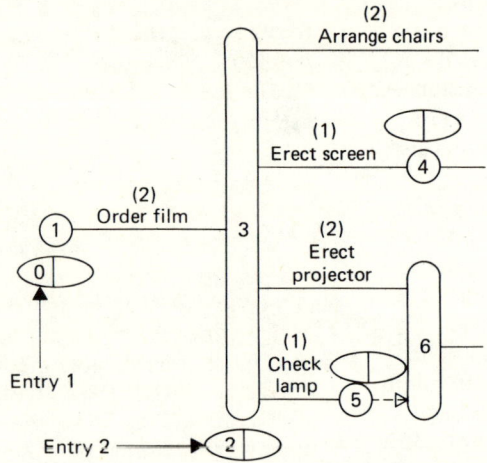

Fig. 8. *The first two entries made showing progressive addition to establish critical path. Since the film ordering is estimated to take 10 minutes and each unit used is five minutes, the unit total, 2, is entered on the left side of the oval at the foot of node 3 (entry 2).*

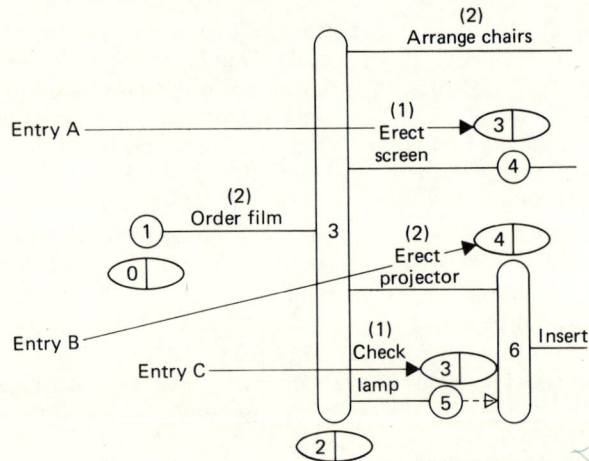

Fig. 9. *Continuation of the progressive addition. The number of five-minute units taken to carry out each activity is shown above each activity.*

2 units and enter the total in the left-hand side of the oval above node 6;

Entry C add the time taken by activity 3–5, 1 unit, and enter the sum in the left-hand side of the oval above node number 5.

The next time oval to be filled in is at the foot of the elongated node 8. The convention for such a node is to insert the largest total time from the lines of parallel activities leading to the node. The totals for the parallel series of activities are the total at the foot of node 3 (2 units) plus:

activity 3–8, *2 units, total 4 units*
activities, erect screen, 3–4, and check screen, 4–8, *2 units, total 4 units*
activities, erect projector, 3–6, and insert film, 6–8, *4 units, total 6 units*
activities, check lamp, 3–5, insert film, 6–8, *3 units, total 5 units*

The largest total is 6 units and this is inserted in the left-hand side of the oval at the foot of node number 8 (figure 10).

To this sum is added the time taken by activity,

8–9, seat audience, 2 units, and the total is entered in the left-hand side of the oval below node 9. The time taken by activity, 9–12, show film, 4 units, is added and the sum, 12 units, is entered in the left-hand side of the oval at the foot of node 12.

In the last section of the network there are again parallel series of activities. The largest total is entered in the oval at the foot of the final elongated node 16.

The diagram in figure 10 shows the result of the progressive addition. The next stage is to proceed back through the network but in this stage the time taken by each activity is progressively substracted from the total time. When there is a choice, where activities are carried out in parallel, the lowest figure is taken.

The final total is entered on the right-hand side of the oval at the foot of node 16. Since each of the activities, 13–15 and 14–16 are 'dummy' activities which use no resources, including time, the entry on the right-hand side of the ovals above nodes 13 and 14 will be the same as the final total, 14 units of time.

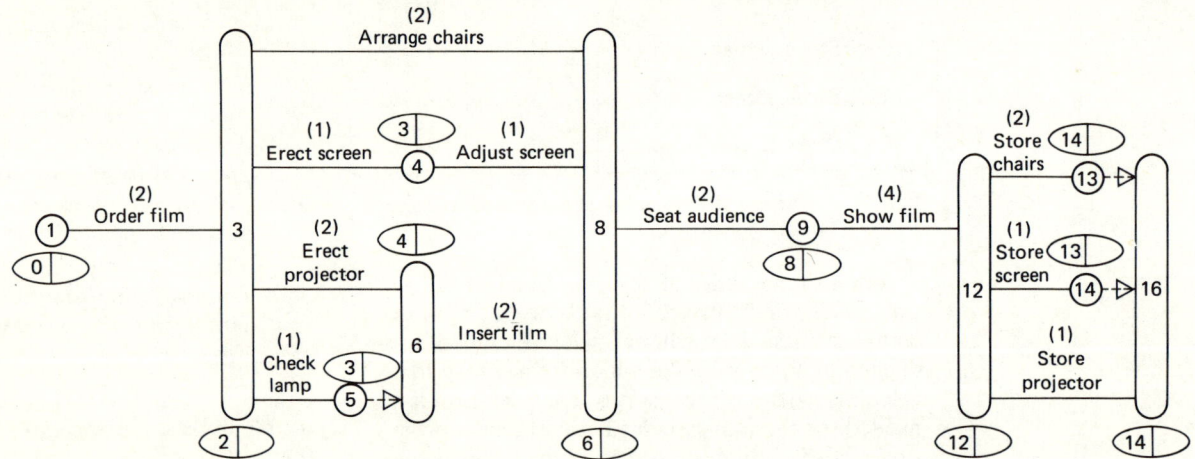

Fig. 10. *Completion of progressive addition showing that the total time taken by the project is 14 units of time (or 70 minutes). All entries have been made on the left-hand side of each oval.*

To find the critical path it is necessary now to subtract each activity time regressively, this time selecting the shortest activity times at each elongated node.

To find the time to be entered in the right-hand side of the oval at the foot of node number 12 we must select the smallest of the times left after subtracting the parallel activity times thus:

14 minus 2 = *12 time units* (14 minus 2, store chairs, activity 12–13)
14 minus 1 = *13 time units* (14 minus 1, store screen, activity 12–14)
14 minus 1 = *13 time units* (14 minus 1, store projector, activity 12–16)

Since the time after activity, store chairs, 12–13, is the least − 12 units, this figure is entered in the right-hand side of the oval at the foot of node number 12 (see figure 11).

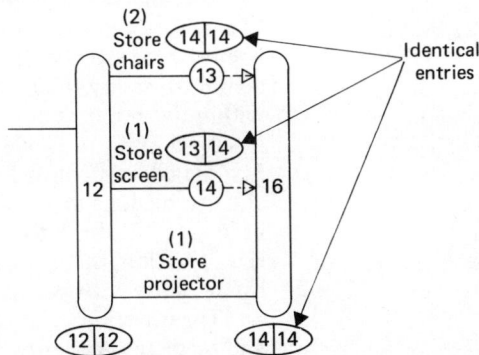

Fig. 11. *The start of regressive subtraction of activity times. Where there are parallel activities the smallest total is entered on the right-hand side of each oval.*

From the total, 12 units, is substracted the time taken to show the film, 4 units, and the time left, 8 units, is inserted in the right-hand side of the oval below node number 9.

The time taken to seat the audience, 2 units, is subtracted from the time in the oval below node 9,

and the result is inserted in the right-hand side of the oval below the elongated node number 8, 6 units. From this total, 6 units, is subtracted the 1 unit taken to check the screen, and the result, 5 units, is entered in the right-hand side of the oval above node 4.

The time taken to insert the film, 2 units, is also subtracted from the 6 units. The result is entered in the right-hand side of the oval above node 6. Since there are a series of parallel activities it is necessary to select the smallest figure to enter in the oval at the foot of node 3.

The last stage of the regressive subtraction is to take from the 2 units left in the oval at the foot of node 3 the time taken ordering the film. This leaves no time at all. This completes the regressive subtraction process and the result should be as in the diagram in figure 12.

Identification of the critical path. When the progressive addition of the activity times and the regressive subtraction is completed and entered in the ovals, it can be seen that in some ovals the two times are identical, and in others the two times

differ. Where the two figures differ the difference is the amount of spare time or 'slack'. A delay in these activities within the slack would not increase the total time taken by the project. In other ovals the addition and subtraction figures are the same. These are the 'critical' activities. There is no slack time for these activities. Any delay in completing these activities would delay the whole project. These critical activities are usually marked by dotted or coloured lines to denote their importance. They form the critical path through the whole project. Any delay in completing this critical path would increase the total time taken to complete the whole project. This is why the technique is often called critical path analysis or critical path method (see figure 13).

By using more efficient equipment, more manpower or more equipment, the activities along the critical path may be completed in less time. This would then reduce the total project time. Care has to be taken however to ensure that when the time taken by activities is shortened along the critical path, other series of activities do not then become 'critical' and cause a hold-up.

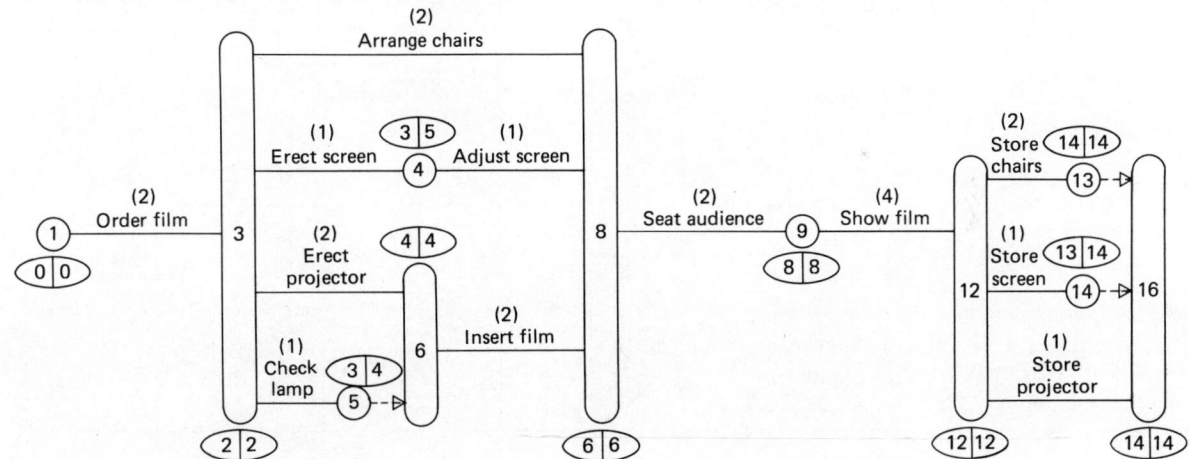

Fig. 12. *Completion of regressive subtraction.*

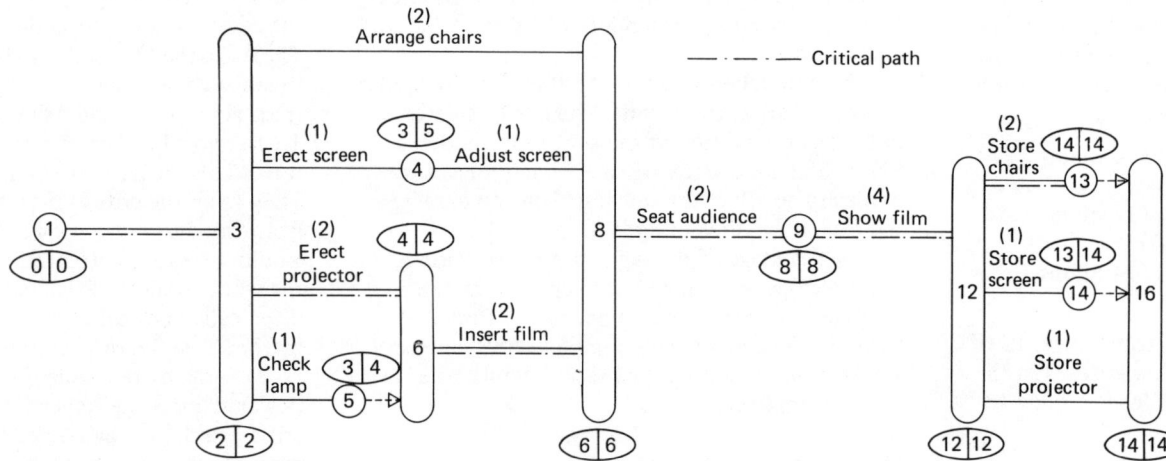

Fig. 13. *The series of activities, the critical activities, identified and shown by dotted lines. This is called the critical path.*

Table of floats — showing film in school

	Activity	Activity number	Time in minutes												
				5	10	15	20	25	30	35	40	45	50	55	60	65	70
1	Order film	1–3	10												
2	Arrange chairs	3–8	10												
3	Erect screen	3–4	5												
4	Erect projector	3–6	10												
5	Adjust screen	4–8	5												
6	Check spare lamp	3–5	5												
7	Insert film	6–8	10												
8	Seat audience	8–9	10			A									
9	Show film	9–12	20					B							
10	Store chairs	12–13	10										C		
11	Store screen	12–14	5										D		
12	Store projector	12–15	5										E		

Fig. 14. *The arrows show the earliest possible starting time and the latest possible finishing time. The dotted lines indicate 'float' or spare time.*

Table of floats. After the consideration of the time consumed by the project and the identification of the critical path the next step is to construct a table of floats. The reason it is called this will quickly become clear.

Here the activities are listed, with the time they take, the earliest possible starting time and the latest possible finishing time. This forms a bar chart as shown in figure 14.

In the table the critical path is shown as a double line. The various possibilities are shown as dotted lines and an arrow shows the starting and finishing point of each possible variation.

1 The arrangement of the chairs can take place at any time in the twenty minutes from the start of the operation, up to the arrival of the audience. Since it takes ten minutes there is a ten-minute float of spare time, or slack.

2 In the same way the screen takes five minutes to erect, and five to adjust, so that as long as the five-minute allowance for adjustment of the screen is subtracted from the starting time, the screen can be put up at any time after the start, and up to within five minutes of the arrival of the audience, as shown by the dotted line and arrow.

3 The adjustment of the screen takes five minutes and can be done at any time after the screen has been erected. If the screen is put up at the start of the operation, there is a ten-minute float, as shown by the dotted arrow.

4 The insertion of the film is on the critical path and so there is no spare time or float.

5 Checking that there is a spare lamp takes five minutes and this can be done at any time after the start and right up to the seating of the audience. The dotted line showing the period of float is extended backwards to show that checking the existence of spare lamps can take place at any time at all. The checking should take place well before the showing of the film to allow time for the purchase of spares. There should therefore be separate lines of activities; order spare lamps and store in a convenient place: then just before a film-showing,

check that a spare is ready to hand in the projector box.

6   There is no spare time and so no float involved in the insertion of the film as this is one of the critical activities.

7   There is a possibility of a saving of time if the audience is being seated while the film is being inserted and checked. This is shown by a dotted arrow up to the letter A. This might not be wise if wires have to be trailed over a floor. In theory it is possible but the feasibility would depend upon such factors as the availability of members of staff to supervise the quiet entry and seating of the pupils.

8   Note that although the seating of the audience is on the critical path, if the audience is seated while the film is being inserted then it will no longer be one of the critical activities. If the seating of the audience takes place while the film is being inserted, then the next activity can be moved forward as shown by the dotted line terminating in the letter B. The last three activities would also be moved as shown by the dotted lines terminating in letters C, D and E.

9   The showing of the film is one of the critical activities and so there is no float of spare time.

10   There is spare time along the line of storing the screen and storing the projector. It can be seen that since each takes five minutes and they take place in parallel while the chairs are being stored, they could take place in succession instead of in parallel with no extra time required. In other words these activities can be done by the same person, and not two acting separately.

For the sake of simplicity the network given as an example has been curtailed. A number of activities would normally take place before a film-showing. For example there might be a meeting of school staff to decide which films are needed to support and enlarge the teaching programme. Before this a list of available films would be needed from which to select the films most suitable. Then the films would have to be ordered and those actually available listed, and the list circulated to the staff giving the dates for which films have been booked.

In later chapters of the book further examples of tables of floats will be given involving more complex series of activities.

Allocation of resources – manpower

Once the network has been drawn with time allotted to each activity and the critical path established, the deployment of other resources can be decided upon. One of the benefits of this sort of planning is that these decisions can be made during objective discussions. One way of shortening any task, of course, is to employ more people to carry it out, up to a point!

If full use is to be made of audio-visual aids then a part of a school might be made available for this use all the time. In that case the time taken and manpower needed for the arrangement of chairs and erection of apparatus is minimised.

Allocation of resources – equipment

After the analysis of the time taken to complete a project has been made, and the number of people to be employed is being considered then the deployment of new equipment may be also considered.

In the case of film-showing one of the bottlenecks might be the lack of sufficient skilled operators. In that case it might be worth considering buying a self-threading projector which could be used by unskilled operators. This might also cut down the time taken to insert the film by skilled as well as unskilled operators.

The other item of equipment which might be altered could be the screen. Even if it was not thought worth while using one room as an audio-visual room, if a different type of screen was used it could save time. If a screen is fastened to a ceiling then it need only be pulled down to enable it to be used. This would not then have to be taken from store, erected and then returned.

These simple examples of the way in which the technique of network analysis can be used in order to weigh up the merits of the economical use of personnel, their time, and the best use of equipment should point the way for a similar consideration of more complex projects.

The way in which time is allocated, and a table of floats showing the amount of slack, or spare time, is used has been demonstrated in a simple form, which can be extended into the more complicated.

No one would use the technique on such a simple project, but having shown the principles in simple form, these can now be extended into the more complex. The same sort of planning does of course take place informally as a matter of common sense in the interests of efficiency. In effect this technique is a reduction of informal common sense, to a disciplined systematic process.

An attempt at some of the following problems might help make the technique clearer. These are all based on situations which arise in schools and for which some sort of planning is required, but not perhaps the application of network analysis. Nevertheless, they may serve as additional demonstrations that the technique is the systematic application of common sense.

It is not necessary for the reader to tackle all the problems. They are designed to enable him to become familiar with the various aspects of the technique.

The problems set are planning and organising:

1   A school medical examination.
2   Fire precautions and fire drill.
3   Accident and first-aid procedure in school.
4   The organisation of school football or netball matches.
5   The annual general meeting of a parent–teacher association.

Even if no attempt is made to solve the problems, the networks may prove to be of interest to those who are involved in such activities, or they may be used to show teachers and others what activities are involved in these events. Solutions are suggested on pp. 112–18.

EXERCISES

Problem 1. *School medical examination*

Activity

1 Receipt of and noting the date of the proposed examination.
2 Writing for record cards of pupils newly entered the school.
3 Confirmation of the date, and ascertainment of the requirements of the medical staff of meals and refreshments.

Arrange these in a logical sequence ready for insertion in a network analysis.
If the allocations of time for the above activities are as follows, insert them in time ovals. (Use ¼-hour time units.)

1 Receipt of date, etc. – ½ hour
2 Writing for record cards – 2 hours
3 Confirmation of date, etc. – 1 hour

Problem 2. *School fire precautions and fire drill*

Draw the first part of a network analysis of the organisation of a school fire precautions and fire drill.

Activity	Time (¼-hour units)
1 Read relevant education authority regulations	1
2 Seek advice of fire officer	2
3 Check hospital facilities, police phone	1
4 Check school first-aid routines	1
5 Seek advice of local inspectors	1

In drawing the network consider: how many activities can take place in parallel; how many must follow in sequence; which must take place first, and which must follow.

Problem 3

Try to continue the network analysis of the organisation of a school medical examination given the following activities and times.

Activity	Time (¼-hour units)
1 Write for medical cards for new pupils	8
2 Note date of examination	2
3 Confirm date, etc.	4
4 Check medical cards	4
5 Notify parents of date	6
6 Check replies	2
7 Replace, or renew medical cards as necessary	24
8 Allot time for each pupil/parent to see doctor	3
9 Duplicate appointment list	4
10 Check medical room	4
11 Arrange table and chairs in medical room	2
12 Notify caretaker of date and duties	2
13 Arrange for meal and refreshments for doctor	1
14 Send copy of appointment list to doctor	1
15 Send copy of appointment list to teaching staff	1
16 Send copy of appointment list to parents	1
17 Check medical room	1
18 Medical inspection	48
19 Check chairs for waiting parents	1
20 Store table and chairs	2
21 Clear medical room	2
22 File medical cards	6
23 Note suggestions, file notes	4
24 Collect payment for meals and refreshments	1
25 Arrange appointments at clinic, dentist, etc.	4

Problem 4. *A network analysis of school fire precautions and fire drill*

Given that the following is the activity list, draw a network analysis. Use the list of times to find the critical path.

Activity	Time (¼-hour units)
1 Read relevant regulations of education authority	1
2 Seek advice of fire officer	2
3 Seek advice of local inspectors	1
4 Check school first aid routines	1
5 Check hospital facilities, police phone number	1
6 Obtain first-aid materials	4
7 Arrange member of staff responsible for first-aid	3
8 Ensure attendance of staff on first-aid courses	4
9 Plan contingency routes for classes	4
10 Arrange school assembly point in drills	1
11 Arrange class assembly points	1
12 Arrange method of checking safety of pupils	2
13 Check fire doors	2
14 List danger points	1
15 Check hand rails, etc.	1
16 Check stairs	1
17 Staff meeting	4
18 Arrange practice 'fires' at danger points	1
19 Duplicate notes on fire drills	2
20 Circulate notes	1
21 Arrange fire prevention courses for staff	2
22 Inform kitchen and other staff of duties	1
23 Discuss fire/arson prevention with fire officer	3
24 Implement suggestions	2

25	Hold fire drill	1
26	Revise/modify arrangements as necessary	1
27	Record time taken to assemble	1
28	Make full notes on drills	2
29	File notes	1
30	Continue fire drills, etc.	1

Problem 5

Draw up a list of activities, and time estimates for the activities involved in pre-planning for accidents and emergencies in schools. Draw a network analysis.

Problem 6

Draw up a list of activities involved in the organisation of school football/netball matches. Draw a network analysis of the planning involved.

Problem 7

Draw up a list of activities involved in planning parent–teacher association meetings. Draw a network analysis of the planning and organisation involved.

3 · Simplified technique for network design

As far as the teacher is concerned teaching involves deciding:

1. What to teach
2. The order in which to teach
3. The method to be used

In effect in the early stages the teacher tries to give each pupil the requisite practical experiences which will enable him to abstract from the variety of experiences the concepts underlying them. These need to be presented in a logical order, gradually progressing from the very simple to the more complex.

For each of the major concepts there is a recognisable line of progression along which a pupil can proceed at a slow or fast pace depending upon his ability. The following is an attempt to show this clearly and simply using a simplified and flexible diagrammatic layout. Since mathematics has a clearly recognisable logical sequence it might be wise to start with this subject.

Chapter 4 is a full account of a successful application of the technique described to a mathematics curriculum and its administrative consequences. The following pages deal only with the use of some of the techniques of network analysis in the drawing up of a syllabus network.

In effect we are concerned with a progression of complexity of learning. Experiences in the various concepts of mathematics might be arranged as shown in figure 15.

Some of the drawing conventions of network analysis can be applied to this model of mathematical learning.

1. We can use arrows to represent each activity
2. Each arrow can proceed from left to right
3. We can use nodes to indicate the start of each activity
4. We can number the nodes and use the number to refer to an activity

The use of elongated nodes can be amended by drawing them as shown in figure 16. Figure 16 shows how the conventions, which are dealt with in more detail in the preceding chapter, are used. Each number represents an activity. No activity can start until all the activities leading to it have been completed. Either 8 or 9 can start after 5 has been completed; 49 can only start when 9 has been completed.

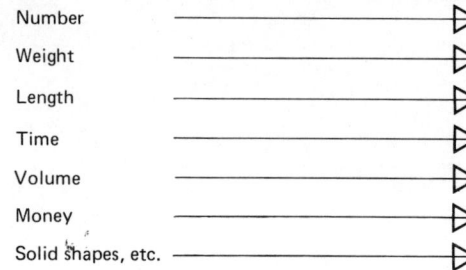

Number ⟶
Weight ⟶
Length ⟶
Time ⟶
Volume ⟶
Money ⟶
Solid shapes, etc. ⟶

Fig. 15. *Mathematical concepts shown as a series of activity arrows.*

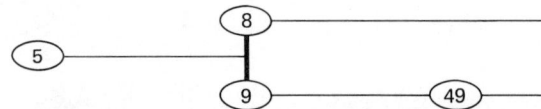

Fig. 16. *Numbering activity arrows. Activities 8 and 9 shown as being dependent upon completion of preceding activity 5.*

A mathematics curriculum design would give a series of activities in columns, each activity having a unique number. Each column would consist of activities at the same level of complexity when thought of in terms of hierarchies of learning. The progression of the columns of activities in order of complexity would be from left to right. (See figure 17.)

In the diagram the first unit of work would consist of activities at the very simplest level of simple discriminations. In this unit activity 1 would be the simple discrimination of number work, for example sorting into heaps of larger and smaller objects.

Activity 2 would be a simple discriminatory activity involving weight, for example weighing two objects in the hands to try to decide which is heavier.

In unit 2 although the activities would involve simple discrimination they would be slightly more difficult. This process of gradual progression of difficulty for each concept would run through the curriculum, unit by unit.

The network upon which the curriculum is based need not be ready to hand for each teacher. Once it has been constructed, critically examined and modified, the results can be typed out as a series of columns of numbered activities for reference.

Drawing such a network is a method of deciding what is to be taught. This is a first decision for any teacher. The next important decision is the order in which the pupils will be presented with the various experiences involved. It is in this area that there are a vast number of theories about hierarchies of learning, some of which conflict. Whichever theory the teacher subscribes to, once the various items in the curriculum are agreed, he has to make the fundamental decision about the order in which the items will be taught. Since the network diagram is a simple way of showing the order in which activities shall be carried out, it ought to be one of the most effective ways of presenting a curriculum to teachers. It should be noted that the diagram in network form can be used equally well to prescribe the course for the individual pupil, as for general curriculum.

Figure 18 is a small section of a network analysis of the mathematics taught to children just starting school.

The concepts involved are listed on the left-hand side of the network and the progression from the simple learning stages to the more complex higher learning hierarchies is from left to right. Each activity is shown as an arrow with a node at the start and has a unique number allotted to it. It is envisaged that a pupil will complete the first column of activities first, then the second column and so on.

The way in which this network was introduced into a school and the building up of an integrated scheme based upon the network is fully described in chapter 4.

Art and craft for young children can be tackled in the same way. At first glance this is not a subject which seems to have a progression susceptible to this type of logical arrangement. However, since many of the activities in art and craft need skills, there should be a progressive build up involved and the pupils should be introduced to simple exercises involving little mental and physical coordination at first and gradually, as the skills develop, more complex and difficult work should be introduced.

In the event the network was used partly as the basis of an art and craft syllabus and partly to help the teachers make full use of the reference books

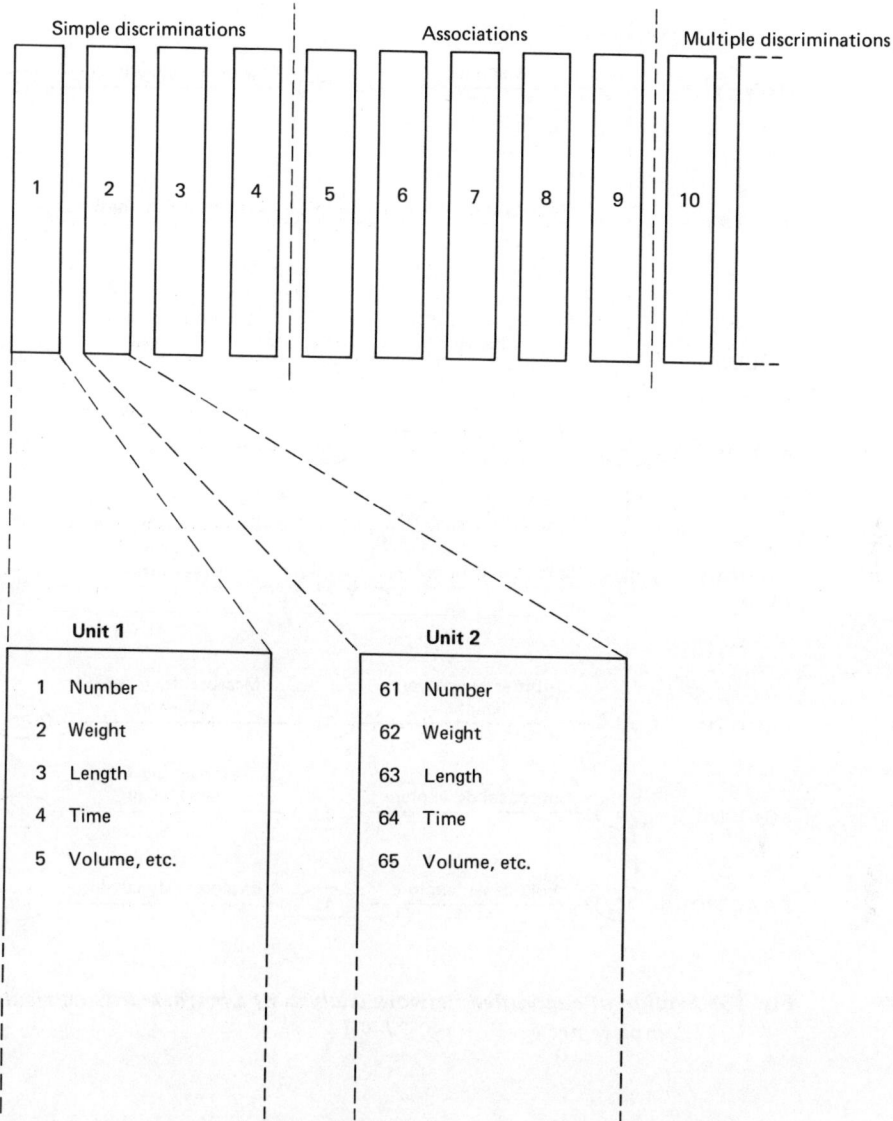

Fig. 17. *Printed layout of a mathematics curriculum for young pupils based upon a network analysis.*

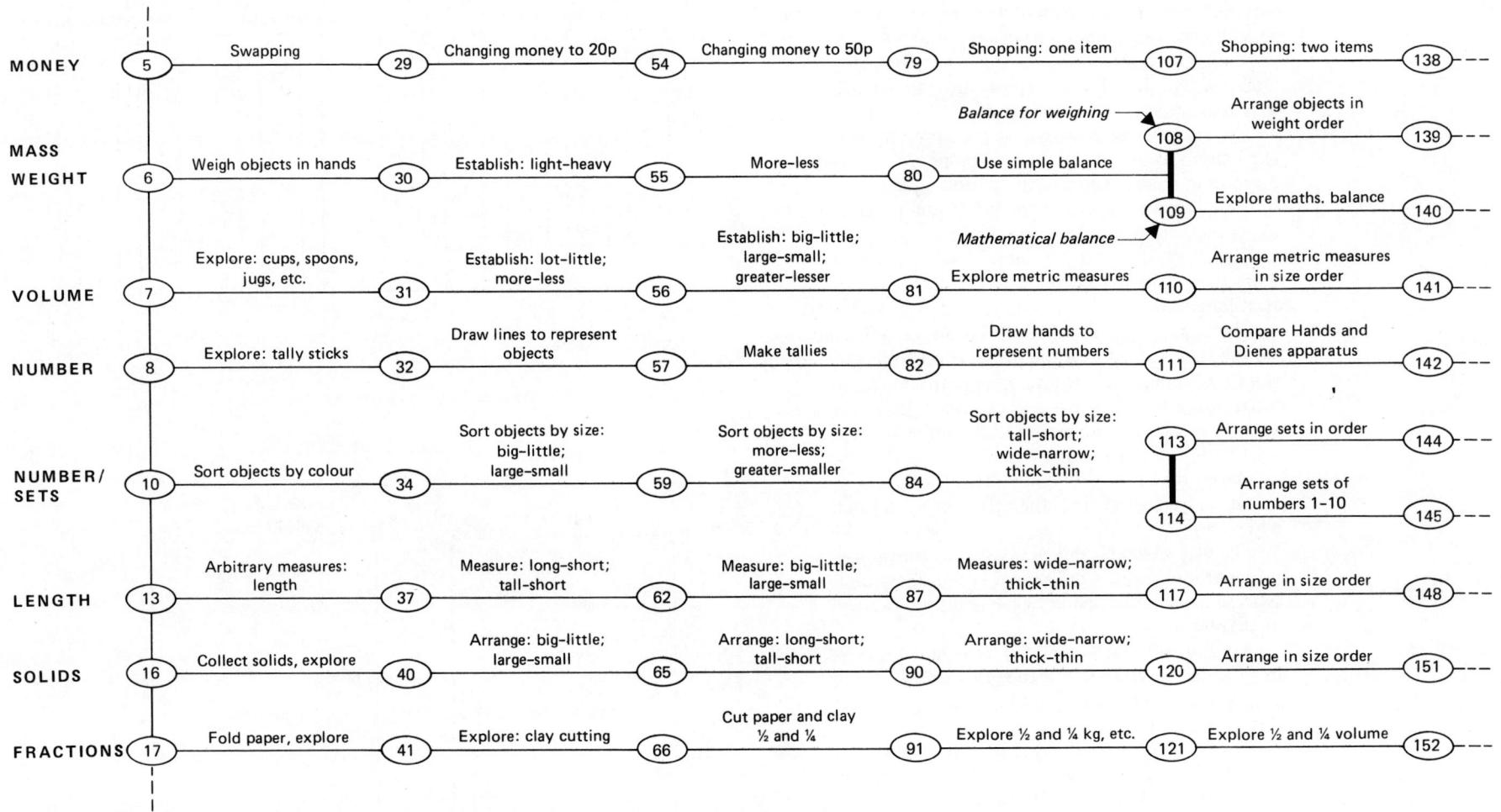

Fig. 18. *Section of a modified network analysis of a mathematics curriculum for young pupils. (Complete network on pp. 24–9.)*

available. Each book was given a number and this was entered in the network on the appropriate activity arrow (figure 40).

The section in figure 19 gives an idea how this works. Once again the progression from simple to complex can be seen, and its application to two different media.

The way in which the art and craft network was constructed and used is fully dealt with in chapter 5.

Science, of course is a subject in which a logical progression is easily found. The total amount of material which could be included in a science syllabus is immense. But if the same principles can be applied, that the pupil needs a series of experiences starting with simple discriminations and proceeding through a hierarchy of difficulty, a network analysis of primary science is also possible and desirable. This is also dealt with in chapter 5. Chapter 6 contains brief accounts of the application of the technique to reading, English, history and geography

The necessity for simple but effective record-keeping has been outlined. Once a curriculum has been designed as a network diagram every activity is numbered, and from this it is a simple matter to design a record card reproducing these numbers. On this card each activity can be marked off as completed to the teacher's satisfaction.

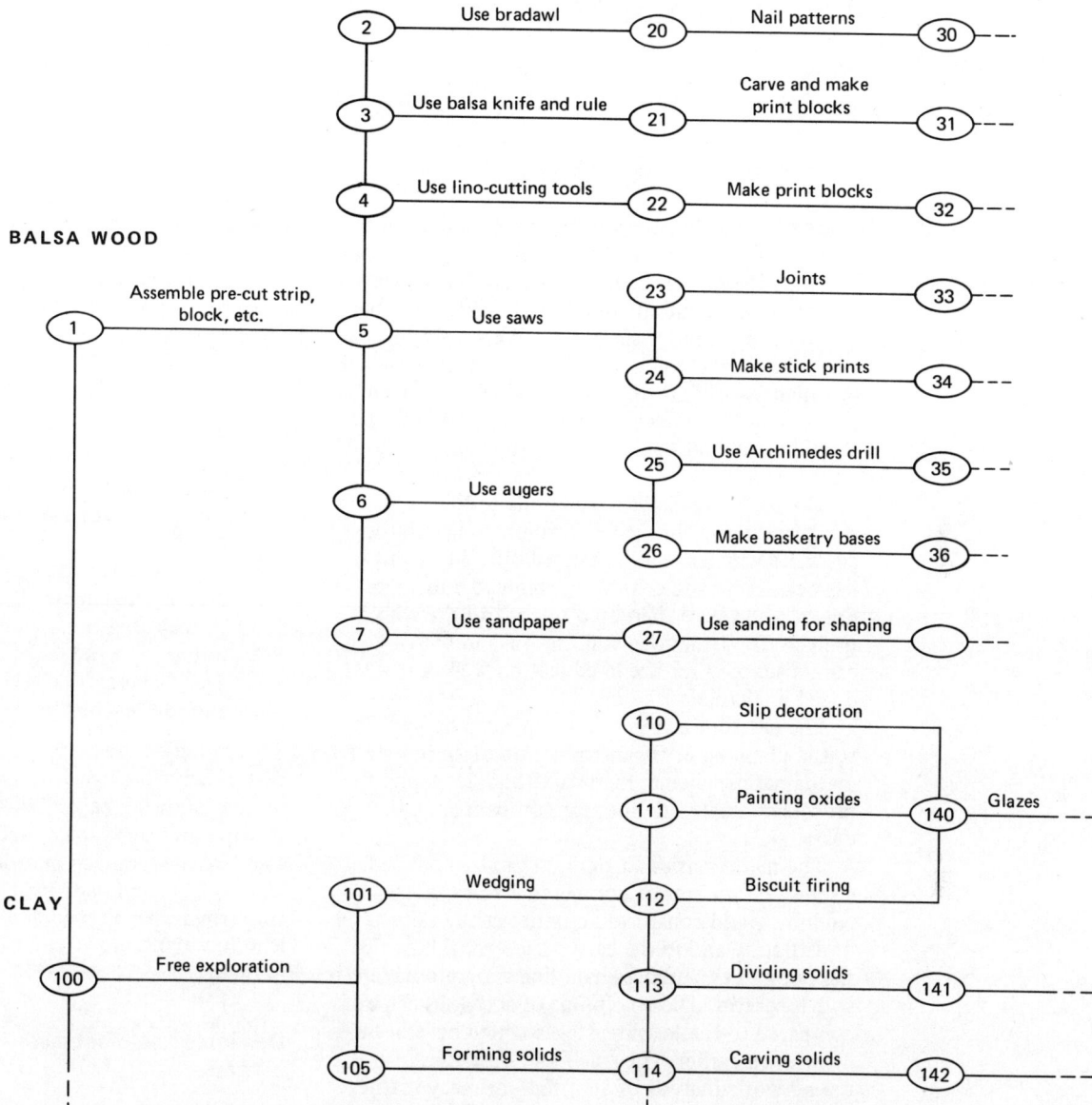

Fig. 19. *A small section of a modified network analysis of an art and craft curriculum for young pupils. The book reference numbers will be found on the complete network (pp. 62–5).*

4 · Mathematics

Some of the problems now facing schools were outlined in chapter 1 and these all affect the teaching of mathematics. In addition there is a shortage of fully trained and qualified mathematics teachers.

To put the matter simply, teachers need to know what they ought to be teaching, and then the order in which to teach it. In any individualised system the problem is to decide what the individual pupil should learn, and then the order in which he should learn.

The network analysis technique follows a similar pattern. The first decision is to list all the activities which are to be carried out. When this has been done and critically examined a diagram is drawn, and this is then used to control the whole project. The technique and the ways in which it can be adapted for use in education were introduced in chapters 2 and 3.

The network analysis technique can be applied to the planning of the introduction of an integrated mathematics scheme. Figures 20 and 21 depict the process as a network and, for comparison, a flow chart.

The mathematics curriculum can be depicted as a series of columns of numbered arrows. Each column would consist of activities at the same level of difficulty and in the early stages would involve the pupil in experiences handling concrete materials and apparatus. These columns of activities can be compared to the learning stages which have been mentioned earlier. The small section taken from the network analysis of an infant–primary mathematics syllabus will serve as an example (figure 22*a*). This shows the practical experiences involving first simple discriminations between a large and a small

object (activities 59 through to 88). The next column (activities 113 to 118) involves multiple discriminations and sorting into order.

In the development of the mathematics scheme the process was as follows:

1   All the activities envisaged as being part of a modern primary mathematics syllabus were listed.
2   This list was modified and amended following suggestions by teachers, head teachers, mathematics students, lecturers, and educational psychologists.
3   The lists were arranged in a logical sequence.
4   A simplified arrow diagram was drawn similar to that used in network analysis, and this was critically examined.
5   This diagram was used to number and then allocate teaching materials.

A modified network analysis of a mathematics curriculum from age 5–13 is shown in figure 22*b* on the following pages (24–9).

Once an agreed list of activities is produced these can be arranged in order of difficulty from left to right. In a mathematics curriculum we are concerned with a range of concepts; number, time, length, weight, etc. These can be arranged as shown in figure 23.

Developing an infant mathematics curriculum network

It was thought at first that a copy of the network drawing of the mathematics curriculum would be of value and interest to staff and pupils and that a

copy might be displayed in the classroom. Since the network was in the early stages of development and was large it was often referred to as 'wallpaper'! In the event the pupils after a first slight interest at seeing their whole course in mathematics displayed, soon lost interest. In the same way class teachers soon found that record cards gave them all the day-to-day information they needed. They could use the record cards to guide the pupils and did not need to refer to the network. Since the network of the whole course was so large, it was reduced in size gradually and then given to teachers as a reference to use when needed in the classroom. A typed list of activities was also given each teacher to include in their files of general information. The first few stages of this list of activities are given below (p. 30).

The network is a valuable tool in the stages of planning and organisation. It is an invaluable aid in allocating materials and when modifications are found necessary as a result of test results and other information, but it is of less value in day-to-day work in the classroom.

For the course designer, and the teaching material designer it has continued relevance and is subject to continuous modification and updating.

Staff meeting

Teaching materials

Discuss implications of scheme

Assemble — 21 — List — 34 — Allocate — 45 — Note shortfall — 55 — Buy — design supplementary — 64 — Allocate

Discuss teaching materials — 10

Storage and retrieval

Examine systems — 22 — Design system — 38 — Buy — make storage / Add to storage — filing system — 63 — Modify as needed

Discuss storage and retrieval — 11

Design number system — 44 — 50 — Allocate materials

Discuss use of audio-visual media — 12

Audio-visual aids

Examine — 23 — Collect audio-visual materials — 33 — Acquire supplementary aids and materials — 54 — Modify numbering — 53 — 60 — Teach pupils storage system — 65 — Try out – modify

Discuss pupil training — 13

Start pupil training — 30 — Train pupils to use audio-visual equipment — 42 — Train pupils to use audio-visual materials

Discuss record-keeping — 14

Decide audio-visual areas — 31 — Make black-out carrels, etc. — Allocate other 'areas' — 51 — Implement — 61 — Teach pupils to use 'areas' — 66 — Try out — modify

Record-keeping

Examine systems — 32 — Evaluate — 43 — Design — print record cards — 52 — Try out — 62 — Modify — 67 — Print — duplicate record cards

1 — 20 — 70

Try out whole scheme — limited numbers — 75

Examine need for new audio-visual aids and materials — 80 — Design new materials — 86 — Print and use

Examine catalogues — 87 — Purchase

Evaluate — 85 — Modify record cards — 88 — 95 — Continue to implement — 100

Modify storage and retrieval — 89

Modify audio-visual and other 'areas'

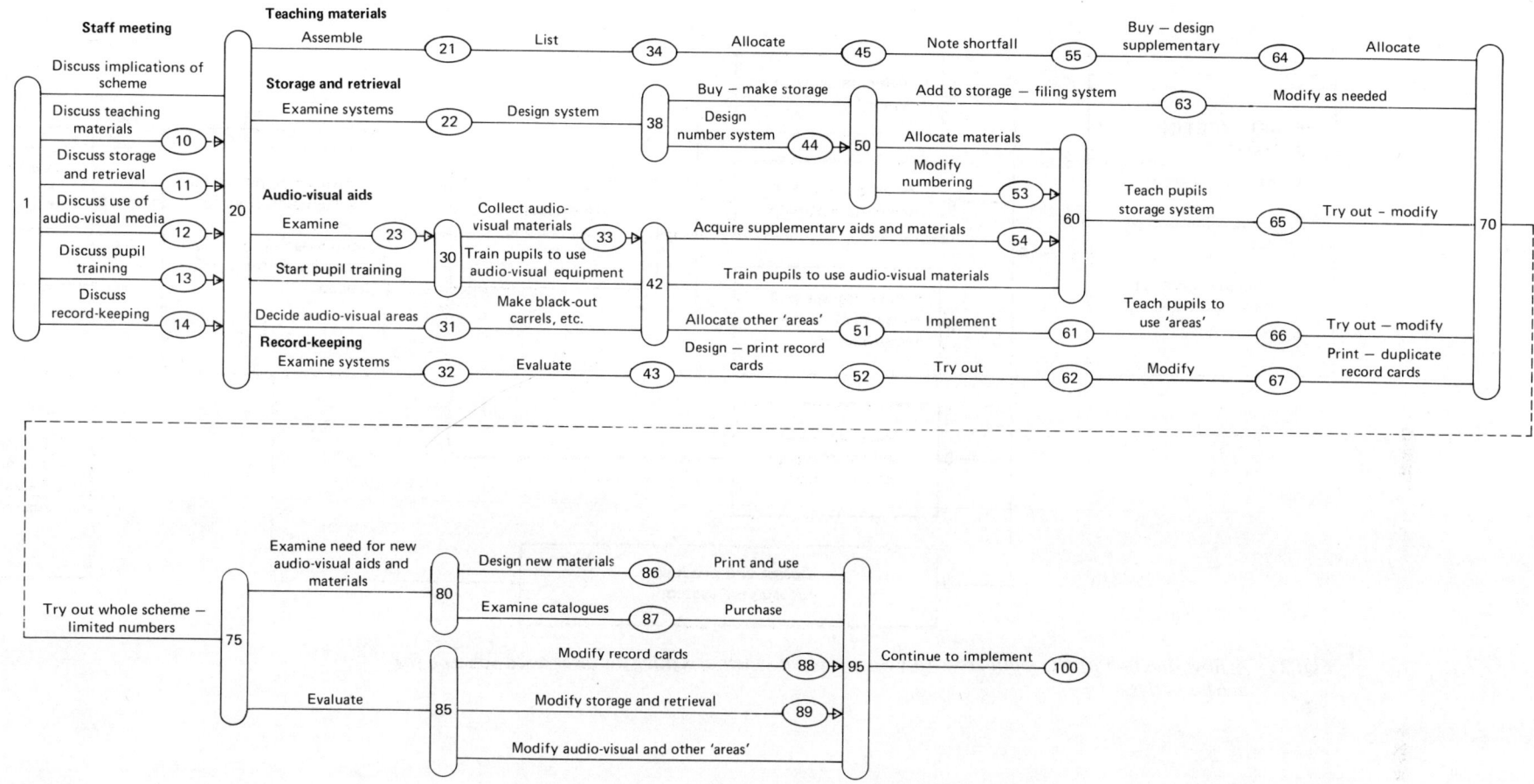

Fig. 20. *A section of a network analysis of the introduction of an integrated mathematics scheme into a school.*

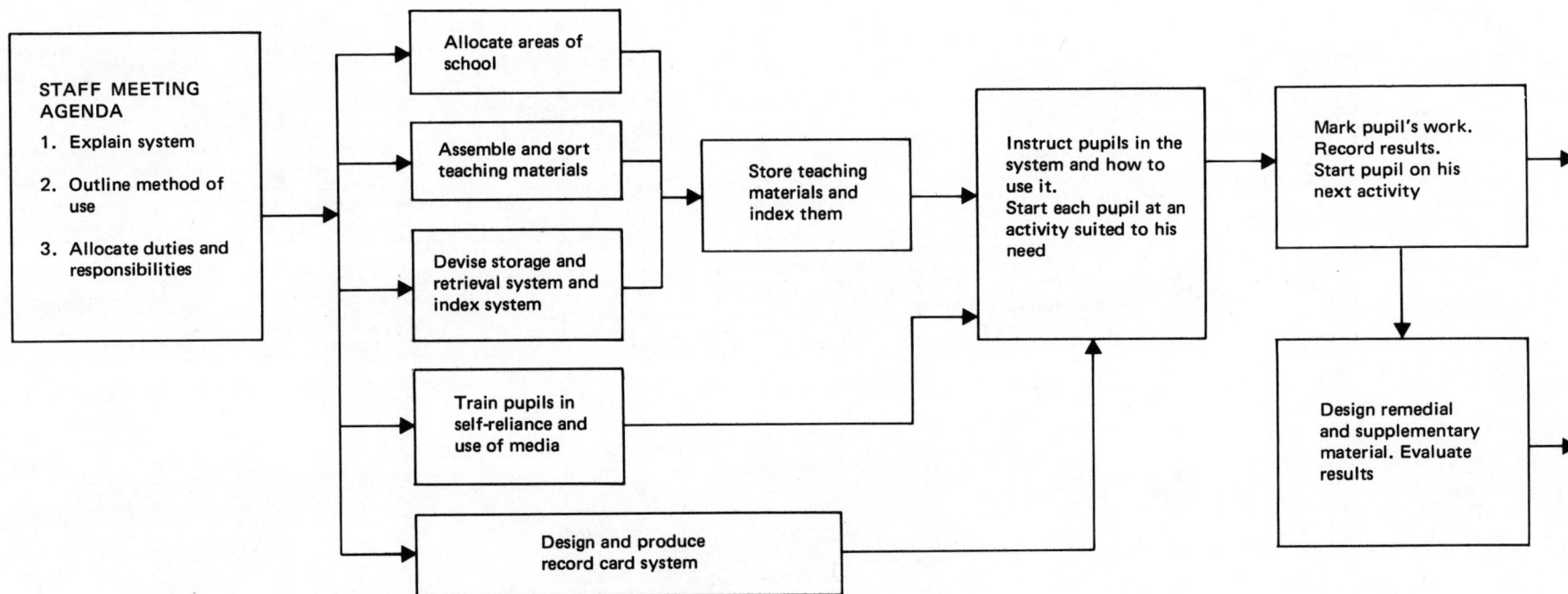

Fig. 21. *A flow diagram of the main areas to be considered when designing an integrated scheme for mathematics.*

Sets

Sort objects into more–less larger–smaller sets

Sort objects into tall–short, wide–narrow, thick–thin

Arrange in order: big–little, large–small

– – – 59 —————— 84 —————— 113 —————————— 144 – – –

114 — Sets of numbers 0–10 — 145 – – –

Geo Board

Make shapes: tall–short

Make shapes: wide–narrow, thick–thin

Make shapes: tall and wide, tall and narrow

– – 60 —————— 85 —————— 115 —————— 146 – – –

Simulation games

Number snap cards

Number Jigsaws

Snakes and ladders

– – 61 —————— 86 —————— 116 —————— 147 – – –

Linear measurement

Using arbitrary measures. Sort into: big–little, large–small

Sort into: long–short, tall–short

Arrange in size order: from big to little and large–small

– – 62 —————— 87 —————— 117 —————— 148 – – –

Structured apparatus

Arrange in sets: big–little, tall–short

Arrange in sets: more–less, larger–smaller

Arrange in size order: big–little, large–small

– – 63 —————— 88 —————— 118 —————— 149 – – –

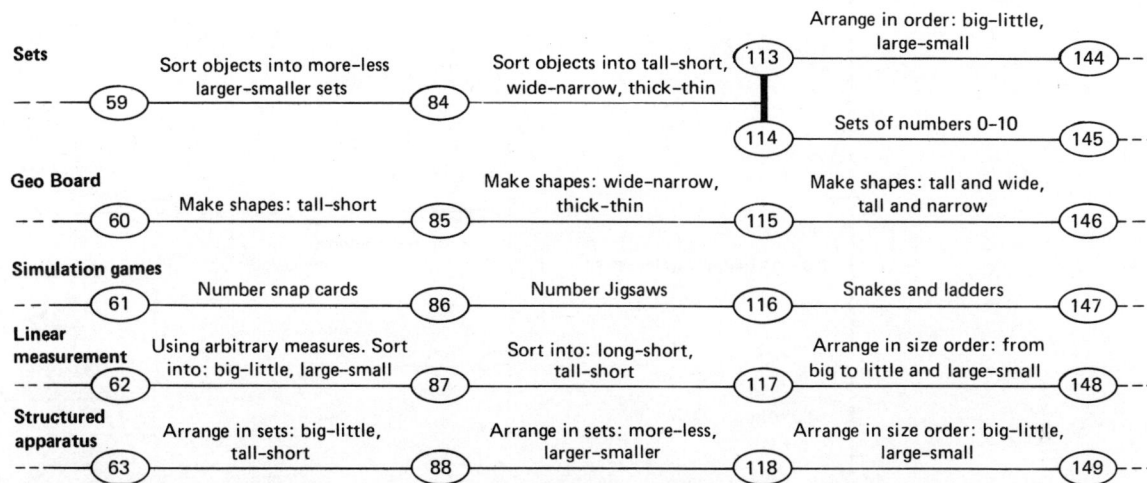

Fig. 22a. *A small section of a modified network analysis of infant-primary mathematics.*

A diagram of the layout of a modified network analysis of a mathematics curriculum to be found on the following pages.

Fig. 22*b*, pt 1 page 24	Fig. 22*b*, pt 2 page 25	Fig. 22*b*, pt 3 page 26
Fig. 22*b*, pt 4 page 27	Fig. 22*b*, pt 5 page 28	Fig. 22*b*, pt 6 page 29

	STAGE 1	STAGE 2	STAGE 3	STAGE 4	STAGE 5	STAGE 6	STAGE 7	STAGE 8	STAGE 9
TIME					*Clock face* (130)	Count hours (160)	Count half hours (190)	Count quarter hours (225)	Count five minutes
(1)	Before–after lunch (25)	Morning–afternoon (50)	Breaks and meals (75)	Explore hour glass (100)	Explore model clock face / *Clock face* (131)	Turn hands–count turns (161)	Count turns and half turns (191)	Count turns and quarter turns (226)	Compare quarter turn and set square
TIME									
(2)	Names of days (26)	Names of months (51)	Days of week (76)	Make–explore sand clock (101)	Make–explore candle clock (132)	Make–explore water clock (162)	Count hours of sunshine (192)	Record hours of sunshine (227)	Measure sun's shadow
(3)	(27)	(52)	(77)	(102)	(133)	(163)	(193)	(228)	
COUNTING FRAME (C/F)					(103) Unifix explore (134)	Unifix explore counting (164)	Unifix explore addition (194)	Unifix explore subtraction (229)	Unifix explore multiplication
(4)	Explore (28)	Use C/F to count (53)	Number recognition (78)	Use C/F to add	(104) Use C/F to subtract (135)	Use C/F to add and subtract (165)	Dienes apparatus: addition (195)	Dienes apparatus: subtraction (230)	Dienes apparatus: addition and subtraction
					(105) Use C/F to multiply (136)	C/F division (166)	Dienes apparatus: multiplication (196)	Dienes apparatus: division (231)	Explore abacus
SHOPPING					(106) Written numbers 0–9 (137)	Written numbers 0–100 (167)	Practise writing numerals (197)	Simple computation (232)	Computation — four rules
(5)	Swapping (29)	Money changing to 20p (54)	Money changing to 50p (79)	Shopping – 1 item (107)	Shopping – 2 items (138)	Shopping – 3 items (168)	Simple bills (198)	Harder bills (233)	Shopping by weight
WEIGHT					*Weighing balance* (108) Arrange objects in weight order (139)	Explore metric weights (169)	Practise using metric weights (199)	Practise increasing discrimination (234)	Practical weighing and computation
(6)	Weigh objects in hands (30)	Weigh objects in hands to establish heavy–light (55)	Weigh objects in hands to establish more–less (80)	Use simple balance		Use M/B to show addition (170)	Use M/B to show subtraction (200)	Use M/B to show multiplication (235)	Use M/B to show division
VOLUME					*Mathematical balance (M/B)* (109) Explore M/B (140)				
(7)	Explore: cups, spoons, jugs, etc. (31)	Establish: lot–little, more–less (56)	Establish: big–little, large–small, greater–lesser (81)	Compare metric measures (110)	Arrange metric measures in size order (141)	Refine metric size ordering (171)	Compare metric weights and volume (201)	Continue comparison (236)	Practical work volume and computation
NUMBER SYSTEMS									
(8)	Tally sticks (32)	Draw lines to represent objects (57)	Make tallies (82)	Draw hands to represent numbers (111)	Compare Hands and Dienes apparatus (142)	Make number line or number ladder (172)	Explore number line or ladder (202)	Make number line to 100 (237)	Use number line to add
(9)	(33)	(58)	(83)	(112)	Explore Roman numerals (143)	Explore Babylonian numerals (173)	Explore Egyptian numerals (203)	Explore Chinese numerals (238)	Explore Greek numerals
SETS					Arrange sets in order: big–little, large–small (144)	Arrange sets in order: more–less, greater–lesser (174)	Arrange sets in order: tall–short, wide–narrow (204)	Arrange sets in order: thick–thin (239)	Arrange: dots–lines in sets
(10)	Sort objects by colour (34)	Sort objects by size: big–little, large–small (59)	Sort objects by size: more–less, greater–lesser (84)	Sort objects by size: tall–short, wide–narrow, thick–thin (113)	(114) Sets of numbers 0–10 (145)	Sets of numbers 0–20 (175)	Arrange numbers in order (205)	Sort into sets by number (240)	Sort into sets of given numbers

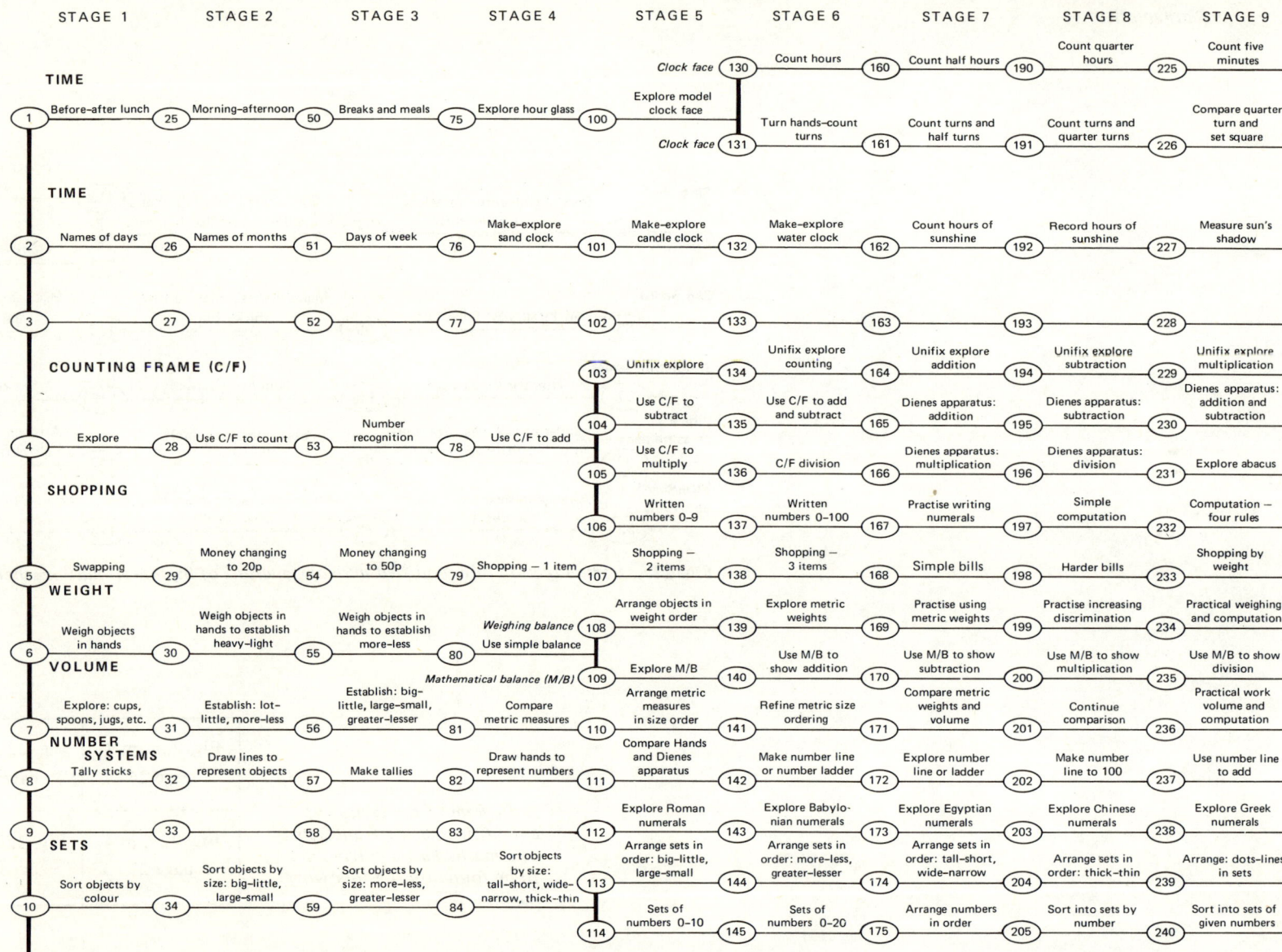

Fig. 22b (part 1). *First of six parts of a modified network analysis of a mathematics curriculum for pupils age 5–13 years.*

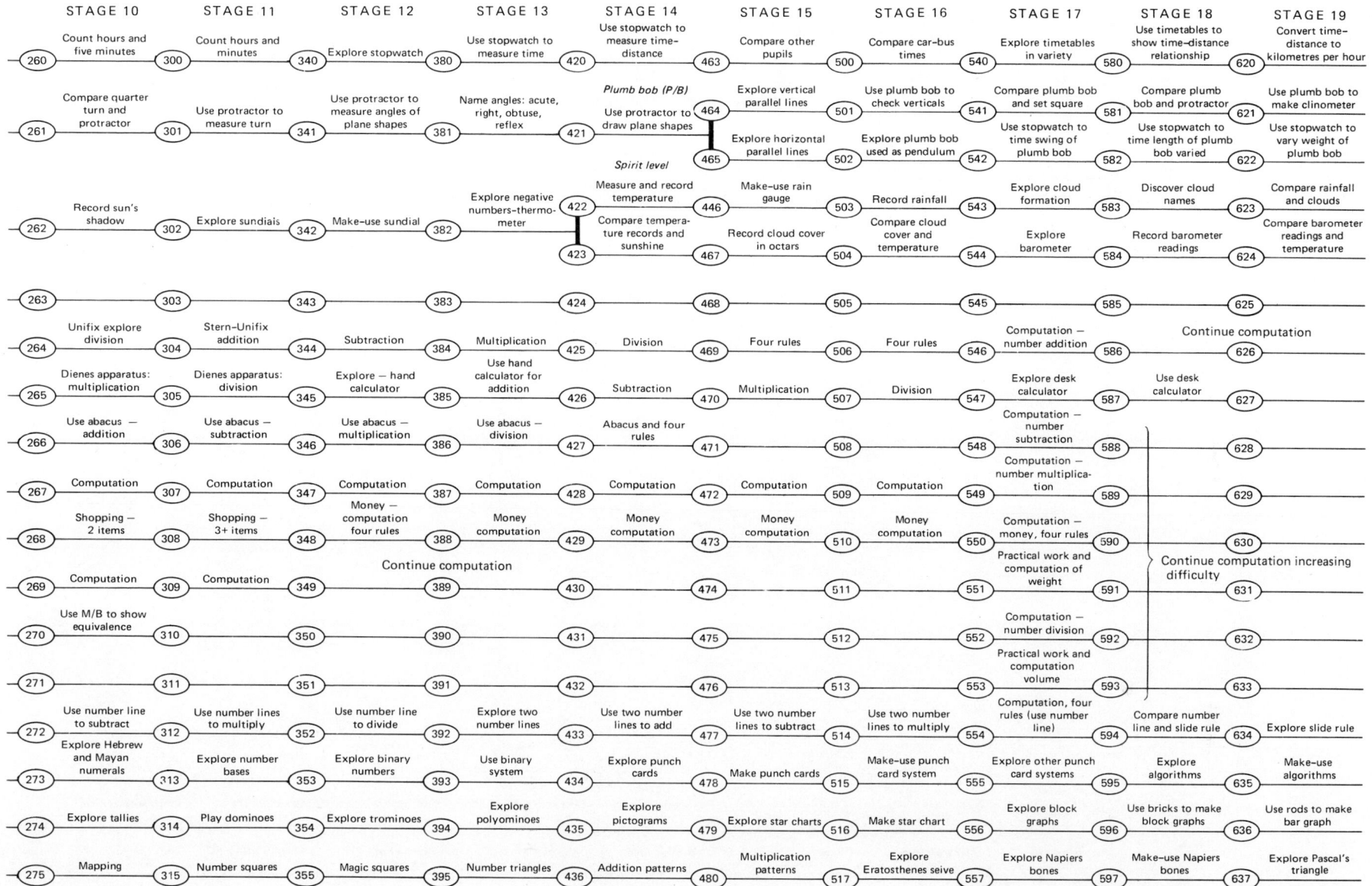

STAGE 10	STAGE 11	STAGE 12	STAGE 13	STAGE 14	STAGE 15	STAGE 16	STAGE 17	STAGE 18	STAGE 19
260 Count hours and five minutes	300 Count hours and minutes	340 Explore stopwatch	380 Use stopwatch to measure time	420 Use stopwatch to measure time–distance · 463	500 Compare other pupils	540 Compare car–bus times	580 Explore timetables in variety	620 Use timetables to show time–distance relationship	Convert time–distance to kilometres per hour
261 Compare quarter turn and protractor	301 Use protractor to measure turn	341 Use protractor to measure angles of plane shapes	381 Name angles: acute, right, obtuse, reflex	421 *Plumb bob (P/B)* · 464 Use protractor to draw plane shapes · 465 *Spirit level*	501 Explore vertical parallel lines / Explore horizontal parallel lines	541 Use plumb bob to check verticals / Explore plumb bob used as pendulum	581 Compare plumb bob and set square / Use stopwatch to time swing of plumb bob	621 Compare plumb bob and protractor / Use stopwatch to time length of plumb bob varied	Use plumb bob to make clinometer / Use stopwatch to vary weight of plumb bob
262 Record sun's shadow	302 Explore sundials	342 Make–use sundial	382 Explore negative numbers–thermometer · 422 / 423	446 Measure and record temperature / 467 Compare temperature records and sunshine	503 Make–use rain gauge / 504 Record cloud cover in octars	543 Record rainfall / 544 Compare cloud cover and temperature	583 Explore cloud formation / 584 Explore barometer	623 Discover cloud names / 624 Record barometer readings	Compare rainfall and clouds / Compare barometer readings and temperature
263	303	343	383	424 · 468	505	545	585	625	
264 Unifix explore division	304 Stern–Unifix addition	344 Subtraction	384 Multiplication	425 Division · 469	506 Four rules	546 Four rules	586 Computation — number addition	626 Continue computation	
265 Dienes apparatus: multiplication	305 Dienes apparatus: division	345 Explore — hand calculator	385 Use hand calculator for addition	426 Subtraction · 470	507 Multiplication	547 Division	587 Explore desk calculator	627 Use desk calculator	
266 Use abacus — addition	306 Use abacus — subtraction	346 Use abacus — multiplication	386 Use abacus — division	427 Abacus and four rules · 471	508	548	588 Computation — number subtraction	628	
267 Computation	307 Computation	347 Computation	387 Computation	428 Computation · 472	509 Computation	549 Computation	589 Computation — number multiplication	629	
268 Shopping — 2 items	308 Shopping — 3+ items	348 Money — computation four rules	388 Money computation	429 Money computation · 473	510 Money computation	550 Money computation	590 Computation — money, four rules	630	
269 Computation	309 Computation	349 Continue computation · 389	430 · 474	511	551 Practical work and computation of weight	591	Continue computation increasing difficulty · 631		
270 Use M/B to show equivalence	310	350	390	431 · 475	512	552 Computation — number division	592	632	
271	311	351	391	432 · 476	513	553 Practical work and computation volume	593	633	
272 Use number line to subtract	312 Use number lines to multiply	352 Use number line to divide	392 Explore two number lines	433 Use two number lines to add · 477	514 Use two number lines to subtract	554 Use two number lines to multiply	594 Computation, four rules (use number line)	634 Compare number line and slide rule	Explore slide rule
273 Explore Hebrew and Mayan numerals	313 Explore number bases	353 Explore binary numbers	393 Use binary system	434 Explore punch cards · 478	515 Make punch cards	555 Make–use punch card system	595 Explore other punch card systems	635 Explore algorithms	Make–use algorithms
274 Explore tallies	314 Play dominoes	354 Explore trominoes	394 Explore polyominoes	435 Explore pictograms · 479	516 Explore star charts	556 Make star chart	596 Explore block graphs	636 Use bricks to make block graphs	Use rods to make bar graph
275 Mapping	315 Number squares	355 Magic squares	395 Number triangles	436 Addition patterns · 480	517 Multiplication patterns	557 Explore Eratosthenes seive	597 Explore Napiers bones	637 Make–use Napiers bones	Explore Pascal's triangle

Fig. 22b (part 2).

25

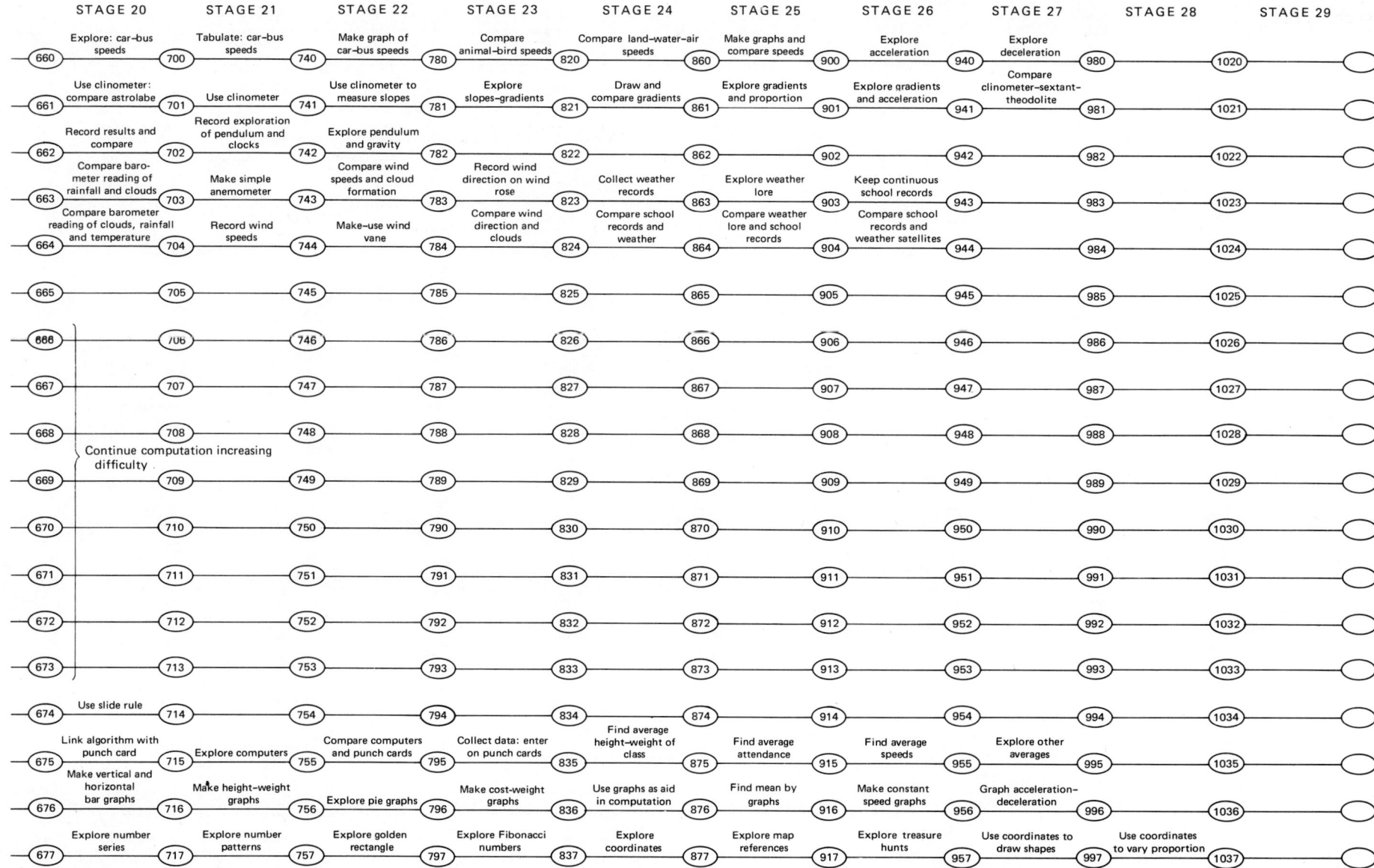

STAGE 20	STAGE 21	STAGE 22	STAGE 23	STAGE 24	STAGE 25	STAGE 26	STAGE 27	STAGE 28	STAGE 29
660 Explore: car-bus speeds	700 Tabulate: car-bus speeds	740 Make graph of car-bus speeds	780 Compare animal-bird speeds	820 Compare land-water-air speeds	860 Make graphs and compare speeds	900 Explore acceleration	940 Explore deceleration	980	1020
661 Use clinometer: compare astrolabe	701 Use clinometer	741 Use clinometer to measure slopes	781 Explore slopes-gradients	821 Draw and compare gradients	861 Explore gradients and proportion	901 Explore gradients and acceleration	941 Compare clinometer-sextant-theodolite	981	1021
662 Record results and compare	702 Record exploration of pendulum and clocks	742 Explore pendulum and gravity	782	822	862	902	942	982	1022
663 Compare barometer reading of rainfall and clouds	703 Make simple anemometer	743 Compare wind speeds and cloud formation	783 Record wind direction on wind rose	823 Collect weather records	863 Explore weather lore	903 Keep continuous school records	943	983	1023
664 Compare barometer reading of clouds, rainfall and temperature	704 Record wind speeds	744 Make-use wind vane	784 Compare wind direction and clouds	824 Compare school records and weather	864 Compare weather lore and school records	904 Compare school records and weather satellites	944	984	1024
665	705	745	785	825	865	905	945	985	1025
666	706	746	786	826	866	906	946	986	1026
667	707	747	787	827	867	907	947	987	1027
668	708	748	788	828	868	908	948	988	1028
669	709	749	789	829	869	909	949	989	1029
670	710	750	790	830	870	910	950	990	1030
671	711	751	791	831	871	911	951	991	1031
672	712	752	792	832	872	912	952	992	1032
673	713	753	793	833	873	913	953	993	1033
674 Use slide rule	714	754	794	834	874	914	954	994	1034
675 Link algorithm with punch card	715 Explore computers	755 Compare computers and punch cards	795 Collect data: enter on punch cards	835 Find average height-weight of class	875 Find average attendance	915 Find average speeds	955 Explore other averages	995	1035
676 Make vertical and horizontal bar graphs	716 Make height-weight graphs	756 Explore pie graphs	796 Make cost-weight graphs	836 Use graphs as aid in computation	876 Find mean by graphs	916 Make constant speed graphs	956 Graph acceleration-deceleration	996	1036
677 Explore number series	717 Explore number patterns	757 Explore golden rectangle	797 Explore Fibonacci numbers	837 Explore coordinates	877 Explore map references	917 Explore treasure hunts	957 Use coordinates to draw shapes	997 Use coordinates to vary proportion	1037

(Bracket spanning nodes 666–673 in Stage 20: "Continue computation increasing difficulty")

Fig. 22b (part 3).

GEOBOARD

- 11 — Explore–make: large-small shapes
- 35 — Explore–make: big-little shapes
- 60 — Explore–make: greater-lesser, tall-short shapes
- 85 — Explore–make: wide-narrow, thick-thin shapes
- 115 — Explore–make: tall and wide, tall and narrow shapes
- 146 — Explore–make: short and narrow, short and wide shapes
- 176 — Explore–make: broad and narrow, broad and wide shapes
- 206 — Horizontal shapes
- 241 — Vertical shapes

SIMULATION GAMING

- 12 — Play snap
- 36 — Play happy families
- 61 — Number snap cards
- 86 — Number jigsaws
- 116 — Play snakes and ladders
- 147 — Play Ludo, etc.
- 177 — Sort animals, etc. into pairs
- 207 — Sort animals into opposites
- 242 — Toss coin — record results

LINEAR MEASUREMENT

- 13 — Hands-feet, etc.
- 37 — Measure: long-short, tall-short
- 62 — Measure: big-little, large-small
- 87 — Measure: wide-narrow, thick-thin
- 117 — Size order: big-little, large-small
- 148 — Size order: wide-narrow, long-short
- 178 — Arrange in height order
- 208 — Arrange in width order
- 243 — Explore pacing stick

STRUCTURED APPARATUS

- 14 — Explore-patterns
- 38 — Arrange: big-little, large-small
- 63 — Arrange: long-short, tall-short
- 88 — Arrange: more-less, larger-smaller
- 118 — Size order: big-little, large-small
- 149 — Size order: wide-narrow, long-short
- 179 — Arrange in height order
- 209 — Arrange in width order
- 244 — Arrange in size order diagonally

DRAWING

- 15 — Draw freehand
- 39 — Draw: straight-crooked, straight-curved
- 64 — Draw: long-short, tall-short
- 89 — Draw: thick-thin, large-small, wide-narrow
- 119 — Use straight edge
- 150 — Use line patterns
- 180 — Use lines to enclose space
- 210 — Trace shapes
- 245 — Curve stitching

SOLID SHAPES

- 16 — Collect-explore
- 40 — Arrange: big-little, large-small
- 65 — Arrange: long-short, tall-short
- 90 — Arrange: wide-narrow, thick-thin
- 120 — Size order: large-small
- 151 — Size order: wide-narrow, long-short
- 181 — Arrange in height order
- 211 — Arrange in width order
- 246 — Arrange sets-round shapes

FRACTIONS

- 17 — Fold paper into half Explore-colour
- 41 — Clay: cut into half, explore
- 66 — Cut paper and clay into half and quarter, explore
- 91 — Explore half and quarter kg, lb and balance
- 121 — Explore: half pint and litre, quarter pint and litre
- 152 — Sort beads, etc. into half and quarter sets
- 182 — Sort structured apparatus
- 212 — Squared paper: structured apparatus, half, quarter, eighth size order
- 247 — Geoboard half, quarter, eighth in size order

PAPER FOLDING

Circular paper (C/P):
- 122 — Explore
- 153 — Fold in half
- 183 — Fold in quarters
- 213 — Fold in segments
- 248 — Join outer points of segments

- 18 — Explore
- 42 — Fold–make: big-little, large-small
- 67 — Fold–make: tall-short long-short
- 92 — Fold–make: wide-narrow, thick-thin
- 125 —
- 156 — Draw irregular shapes, count squares
- 186 — Draw hands, feet - count squares
- 214 — Trace leaves, squares, etc. (Squared paper S/P)
- Draw shapes: same number of squares vertical and horizontal
- 249 — Count squares in shapes

Squared paper (S/P):
- 123 — Make patterns
- 154 — Count squares in patterns
- 184 — Colour shapes, count squares
- 215 — Squared paper (S/P)
- Draw shapes different number of squares vertical and horizontal
- 250 — Count squares in shapes
- 251 — Draw diagonals — colour half

PLANE SHAPES (P/S)

- 19 — Explore P/S
- Arrange sets: big-little
- 68 — Arrange sets: large-small
- 93 — Arrange sets: long-short, tall-short
- 124 — Arrange sets: wide-narrow, thick-thin
- 155 — Arrange sets: similar shape
- 185 — Arrange sets: same number of sides and corners
- 216 — Select sets of four sides, explore
- 252 — Tessellate squares
- 253 — Select triangles and explore
- 254 — Trace triangles and explore
- 255 — Explore round shapes

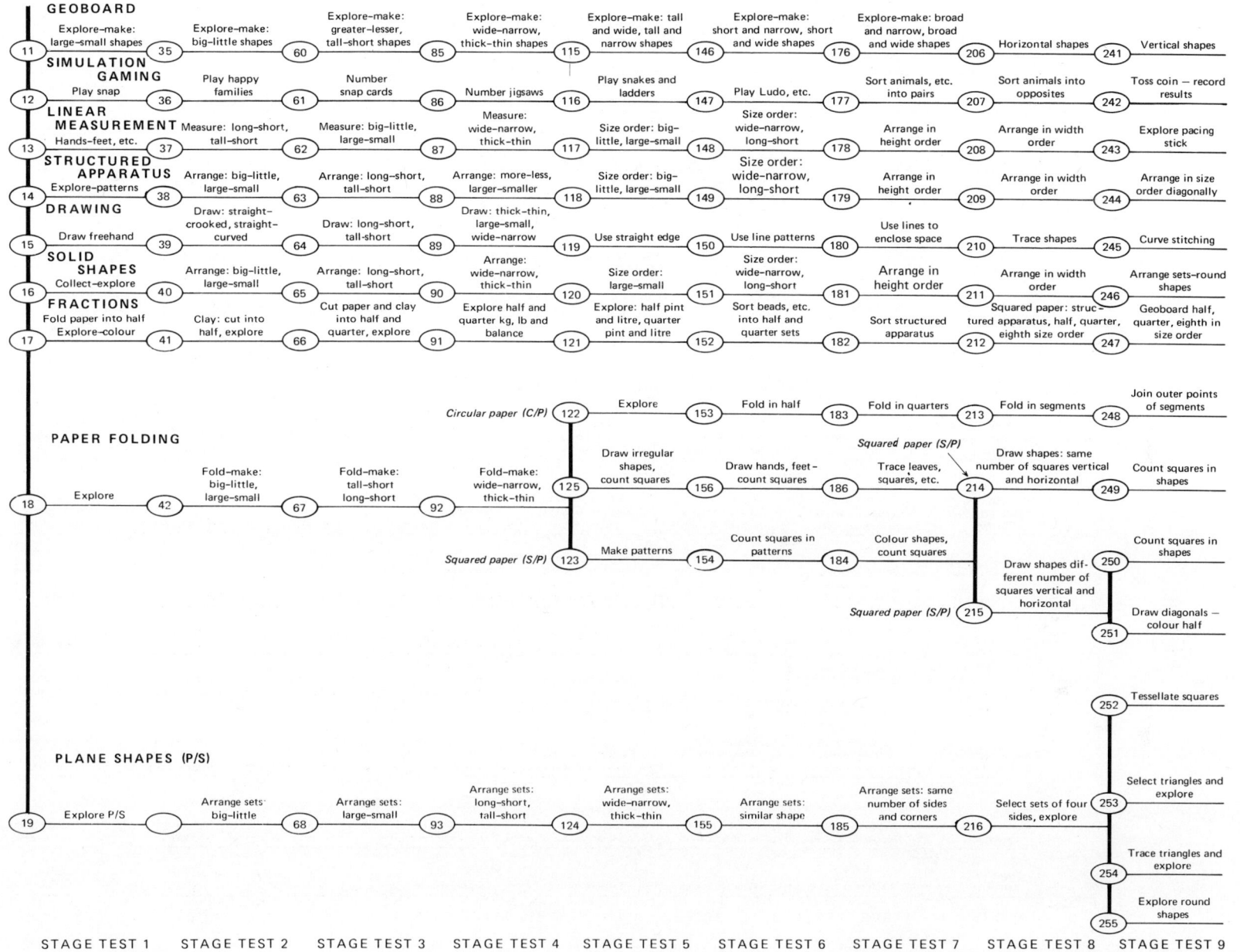

STAGE TEST 1 · STAGE TEST 2 · STAGE TEST 3 · STAGE TEST 4 · STAGE TEST 5 · STAGE TEST 6 · STAGE TEST 7 · STAGE TEST 8 · STAGE TEST 9

Fig. 22b (part 4).

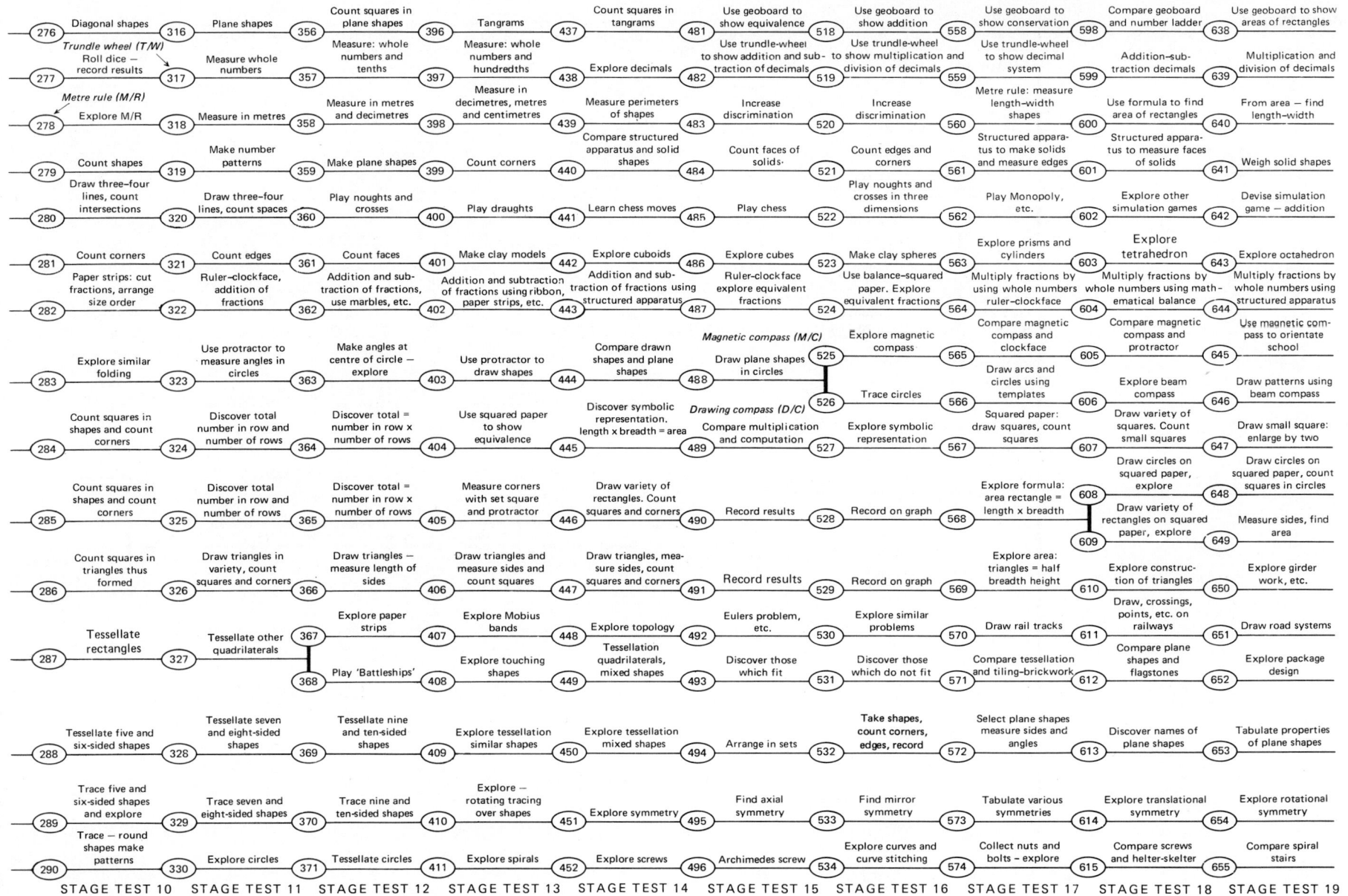

(276)	(316)	(356)	(396)	(437)	(481)	(518)	(558)	(598)	(638)
Diagonal shapes (276)	Plane shapes (316)	Count squares in plane shapes (356)	Tangrams (396)	Count squares in tangrams (437)	Use geoboard to show equivalence (481)	Use geoboard to show addition (518)	Use geoboard to show conservation (558)	Compare geoboard and number ladder (598)	Use geoboard to show areas of rectangles (638)
Trundle wheel (T/W) Roll dice — record results (277)	Measure whole numbers (317)	Measure: whole numbers and tenths (357)	Measure: whole numbers and hundredths (397)	Explore decimals (438)	Use trundle-wheel to show addition and subtraction of decimals (482)	Use trundle-wheel to show multiplication and division of decimals (519)	Use trundle-wheel to show decimal system (559)	Addition–subtraction decimals (599)	Multiplication and division of decimals (639)
Metre rule (M/R) Explore M/R (278)	Measure in metres (318)	Measure in metres and decimetres (358)	Measure in decimetres, metres and centimetres (398)	Measure perimeters of shapes (439)	Increase discrimination (483)	Increase discrimination (520)	Metre rule: measure length–width shapes (560)	Use formula to find area of rectangles (600)	From area — find length–width (640)
Count shapes (279)	Make number patterns (319)	Make plane shapes (359)	Count corners (399)	Compare structured apparatus and solid shapes (440)	Count faces of solids (484)	Count edges and corners (521)	Structured apparatus to make solids and measure edges (561)	Structured apparatus to measure faces of solids (601)	Weigh solid shapes (641)
Draw three–four lines, count intersections (280)	Draw three–four lines, count spaces (320)	Play noughts and crosses (360)	Play draughts (400)	Learn chess moves (441)	Play chess (485)	Play noughts and crosses in three dimensions (522)	Play Monopoly, etc. (562)	Explore other simulation games (602)	Devise simulation game — addition (642)
Count corners (281)	Count edges (321)	Count faces (361)	Make clay models (401)	Explore cuboids (442)	Explore cubes (486)	Make clay spheres (523)	Explore prisms and cylinders (563)	Explore tetrahedron (603)	Explore octahedron (643)
Paper strips: cut fractions, arrange size order (282)	Ruler-clockface, addition of fractions (322)	Addition and subtraction of fractions, use marbles, etc. (362)	Addition and subtraction of fractions using ribbon, paper strips, etc. (402)	Addition and subtraction of fractions using structured apparatus (443)	Ruler-clockface explore equivalent fractions (487)	Use balance-squared paper. Explore equivalent fractions (524)	Multiply fractions by using whole numbers ruler-clockface (564)	Multiply fractions by whole numbers using mathematical balance (604)	Multiply fractions by whole numbers using structured apparatus (644)
Explore similar folding (283)	Use protractor to measure angles in circles (323)	Make angles at centre of circle — explore (363)	Use protractor to draw shapes (403)	Compare drawn shapes and plane shapes (444)	Draw plane shapes in circles (488)	*Magnetic compass (M/C)* (525) / *Drawing compass (D/C)* Trace circles (526)	Explore magnetic compass (565) / Draw arcs and circles using templates (566)	Compare magnetic compass and clockface (605) / Explore beam compass (606)	Use magnetic compass to orientate school (645) / Draw patterns using beam compass (646)
Count squares in shapes and count corners (284)	Discover total number in row and number of rows (324)	Discover total = number in row x number of rows (364)	Use squared paper to show equivalence (404)	Discover symbolic representation. length x breadth = area (445)	Compare multiplication and computation (489)	Explore symbolic representation (527)	Squared paper: draw squares, count squares (567)	Draw variety of squares. Count small squares (607)	Draw small square: enlarge by two (647)
Count squares in shapes and count corners (285)	Discover total number in row and number of rows (325)	Discover total = number in row x number of rows (365)	Measure corners with set square and protractor (405)	Draw variety of rectangles. Count squares and corners (446)	Record results (490)	Record on graph (528)	Explore formula: area rectangle = length x breadth (568)	Draw circles on squared paper, explore (608) / Draw variety of rectangles on squared paper, explore (609)	Draw circles on squared paper, count squares in circles (648) / Measure sides, find area (649)
Count squares in triangles thus formed (286)	Draw triangles in variety, count squares and corners (326)	Draw triangles — measure length of sides (366)	Draw triangles and measure sides and count squares (406)	Draw triangles, measure sides, count squares and corners (447)	Record results (491)	Record on graph (529)	Explore area: triangles = half breadth height (569)	Explore construction of triangles (610)	Explore girder work, etc. (650)
Tessellate rectangles (287)	Tessellate other quadrilaterals (327)	Explore paper strips (367) / Play 'Battleships' (368)	Explore Mobius bands (407) / Explore touching shapes (408)	Explore topology (448) / Tessellation quadrilaterals, mixed shapes (449)	Eulers problem, etc. (492) / Discover those which fit (493)	Explore similar problems (530) / Discover those which do not fit (531)	Draw rail tracks (570) / Compare tessellation and tiling-brickwork (571)	Draw, crossings, points, etc. on railways (611) / Compare plane shapes and flagstones (612)	Draw road systems (651) / Explore package design (652)
Tessellate five and six-sided shapes (288)	Tessellate seven and eight-sided shapes (328)	Tessellate nine and ten-sided shapes (369)	Explore tessellation similar shapes (409)	Explore tessellation mixed shapes (450)	Arrange in sets (494)	Take shapes, count corners, edges, record (532)	Select plane shapes measure sides and angles (572)	Discover names of plane shapes (613)	Tabulate properties of plane shapes (653)
Trace five and six-sided shapes and explore (289)	Trace seven and eight-sided shapes (329)	Trace nine and ten-sided shapes (370)	Explore — rotating tracing over shapes (410)	Explore symmetry (451)	Find axial symmetry (495)	Find mirror symmetry (533)	Tabulate various symmetries (573)	Explore translational symmetry (614)	Explore rotational symmetry (654)
Trace — round shapes make patterns (290)	Explore circles (330)	Tessellate circles (371)	Explore spirals (411)	Explore screws (452)	Archimedes screw (496)	Explore curves and curve stitching (534)	Collect nuts and bolts - explore (574)	Compare screws and helter-skelter (615)	Compare spiral stairs (655)

STAGE TEST 10 STAGE TEST 11 STAGE TEST 12 STAGE TEST 13 STAGE TEST 14 STAGE TEST 15 STAGE TEST 16 STAGE TEST 17 STAGE TEST 18 STAGE TEST 19

28 **Fig. 22*b* (part 5).**

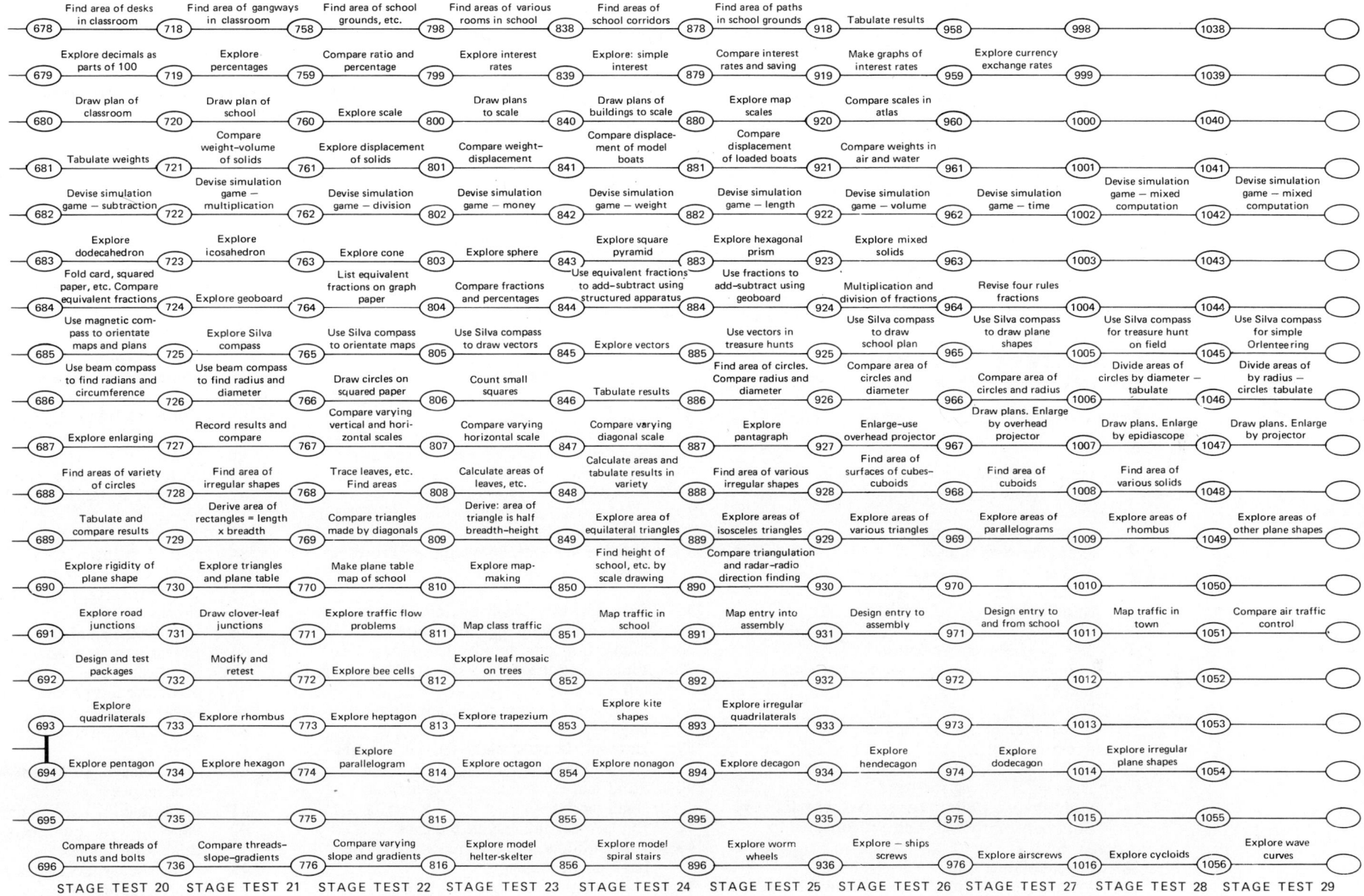

Stage Test 20	Stage Test 21	Stage Test 22	Stage Test 23	Stage Test 24	Stage Test 25	Stage Test 26	Stage Test 27	Stage Test 28	Stage Test 29
678 Find area of desks in classroom	718 Find area of gangways in classroom	758 Find area of school grounds, etc.	798 Find areas of various rooms in school	838 Find areas of school corridors	878 Find area of paths in school grounds	918 Tabulate results	958	998	1038
679 Explore decimals as parts of 100	719 Explore percentages	759 Compare ratio and percentage	799 Explore interest rates	839 Explore: simple interest	879 Compare interest rates and saving	919 Make graphs of interest rates	959 Explore currency exchange rates	999	1039
680 Draw plan of classroom	720 Draw plan of school	760 Explore scale	800 Draw plans to scale	840 Draw plans of buildings to scale	880 Explore map scales	920 Compare scales in atlas	960	1000	1040
681 Tabulate weights	721 Compare weight–volume of solids	761 Explore displacement of solids	801 Compare weight–displacement	841 Compare displacement of model boats	881 Compare displacement of loaded boats	921 Compare weights in air and water	961	1001	1041
682 Devise simulation game — subtraction	722 Devise simulation game — multiplication	762 Devise simulation game — division	802 Devise simulation game — money	842 Devise simulation game — weight	882 Devise simulation game — length	922 Devise simulation game — volume	962 Devise simulation game — time	1002 Devise simulation game — mixed computation	1042 Devise simulation game — mixed computation
683 Explore dodecahedron	723 Explore icosahedron	763 Explore cone	803 Explore sphere	843 Explore square pyramid	883 Explore hexagonal prism	923 Explore mixed solids	963	1003	1043
684 Fold card, squared paper, etc. Compare equivalent fractions	724 Explore geoboard	764 List equivalent fractions on graph paper	804 Compare fractions and percentages	844 Use equivalent fractions to add–subtract using structured apparatus	884 Use fractions to add–subtract using geoboard	924 Multiplication and division of fractions	964 Revise four rules fractions	1004	1044
685 Use magnetic compass to orientate maps and plans	725 Explore Silva compass	765 Use Silva compass to orientate maps	805 Use Silva compass to draw vectors	845 Explore vectors	885 Use vectors in treasure hunts	925 Use Silva compass to draw school plan	965 Use Silva compass to draw plane shapes	1005 Use Silva compass for treasure hunt on field	1045 Use Silva compass for simple Orienteering
686 Use beam compass to find radians and circumference	726 Use beam compass to find radius and diameter	766 Draw circles on squared paper	806 Count small squares	846 Tabulate results	886 Find area of circles. Compare radius and diameter	926 Compare area of circles and diameter	966 Compare area of circles and radius	1006 Divide areas of circles by diameter — tabulate	1046 Divide areas of by radius — circles tabulate
687 Explore enlarging	727 Record results and compare	767 Compare varying vertical and horizontal scales	807 Compare varying horizontal scale	847 Compare varying diagonal scale	887 Explore pantagraph	927 Enlarge–use overhead projector	967 Draw plans. Enlarge by overhead projector	1007 Draw plans. Enlarge by epidiascope	1047 Draw plans. Enlarge by projector
688 Find areas of variety of circles	728 Find area of irregular shapes	768 Trace leaves, etc. Find areas	808 Calculate areas of leaves, etc.	848 Calculate areas and tabulate results in variety	888 Find area of various irregular shapes	928 Find area of surfaces of cubes–cuboids	968 Find area of cuboids	1008 Find area of various solids	1048
689 Tabulate and compare results	729 Derive area of rectangles = length x breadth	769 Compare triangles made by diagonals	809 Derive: area of triangle is half breadth–height	849 Explore area of equilateral triangles	889 Explore areas of isosceles triangles	929 Explore areas of various triangles	969 Explore areas of parallelograms	1009 Explore areas of rhombus	1049 Explore areas of other plane shapes
690 Explore rigidity of plane shape	730 Explore triangles and plane table	770 Make plane table map of school	810 Explore map-making	850 Find height of school, etc. by scale drawing	890 Compare triangulation and radar–radio direction finding	930	970	1010	1050
691 Explore road junctions	731 Draw clover-leaf junctions	771 Explore traffic flow problems	811 Map class traffic	851 Map traffic in school	891 Map entry into assembly	931 Design entry to assembly	971 Design entry to and from school	1011 Map traffic in town	1051 Compare air traffic control
692 Design and test packages	732 Modify and retest	772 Explore bee cells	812 Explore leaf mosaic on trees	852	892	932	972	1012	1052
693 Explore quadrilaterals	733 Explore rhombus	773 Explore heptagon	813 Explore trapezium	853 Explore kite shapes	893 Explore irregular quadrilaterals	933	973	1013	1053
694 Explore pentagon	734 Explore hexagon	774 Explore parallelogram	814 Explore octagon	854 Explore nonagon	894 Explore decagon	934 Explore hendecagon	974 Explore dodecagon	1014 Explore irregular plane shapes	1054
695	735	775	815	855	895	935	975	1015	1055
696 Compare threads of nuts and bolts	736 Compare threads–slope–gradients	776 Compare varying slope and gradients	816 Explore model helter-skelter	856 Explore model spiral stairs	896 Explore worm wheels	936 Explore — ships screws	976 Explore airscrews	1016 Explore cycloids	1056 Explore wave curves

STAGE TEST 20 STAGE TEST 21 STAGE TEST 22 STAGE TEST 23 STAGE TEST 24 STAGE TEST 25 STAGE TEST 26 STAGE TEST 27 STAGE TEST 28 STAGE TEST 29

Fig. 22*b* (part 6).

Simple activities ⟶ Increasingly complex activities ⟶

Stage 1	Stage 2	Stage 3	etc.
1. Number	25.	50.	
2. Time	26.	51.	
3. Length	27.	52.	
4. etc.	28.	53.	
5.	29.	54.	
etc.	etc.	etc.	

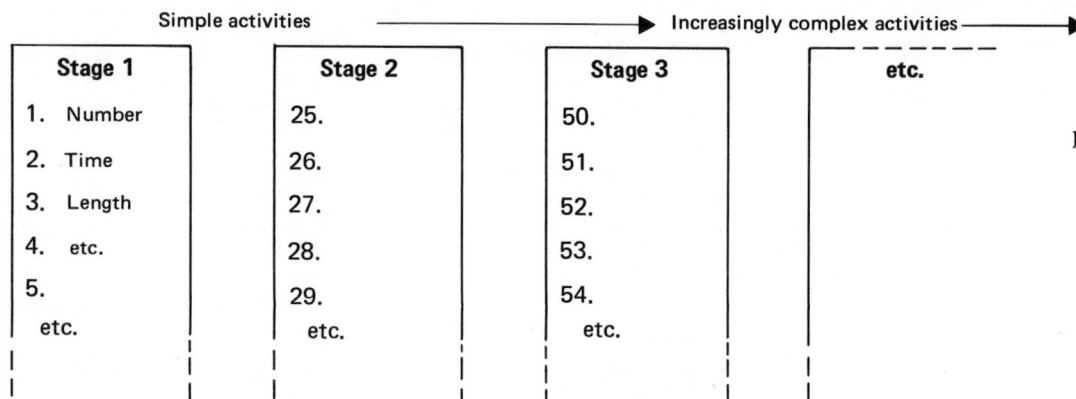

Fig. 23. *A diagram showing how the modified network analysis is laid out for use by teachers. Each stage is intended to cover a series of activities suited to the mathematical development of pupils.*

INFANT-PRIMARY SYLLABUS BASED ON A NETWORK ANALYSIS OF INFANT-PRIMARY MATHEMATICS

Stage 1

Number	Activity
1	Time. Establish before–after lunch
2	Time. Name of days
3	
4	Counting frame. Explore
5	Shopping. Swapping
6	Weight. Weigh objects in hands
7	Volume. Explore cups, spoons, jugs, etc.
8	Number systems. Tally sticks
9	
10	Sets. Sort objects by colour
11	Geoboard. Explore and make small–large shapes
12	Simulation gaming. Play snap
13	Linear measurement. Use hands, feet, paces, explore
14	Structured apparatus. Explore patterns
15	Drawing. Drawing freehand
16	Solid shapes. Collect boxes, tins, etc. Explore
17	Fold paper into half. Colour, continue to explore
18	Paper folding. Explore
19	Plane shapes. Explore box of plane shapes

Stage test

Stage 2

Number	Activity
25	Time. Morning–afternoon
26	Time. Names of the months of the year
27	
28	Counting frame. Use counting frame to count
29	Shopping. Money changing up to 20p.
30	Weight. Weigh objects in hand, establish heavy–light
31	Volume. Arbitrary measures, establish lot–little, more–less
32	Number systems. Draw lines to represent objects
33	
34	Sets. Sort into big–little, large–small
35	Geoboard. Make big–little shapes
36	Simulation gaming. Play happy families
37	Linear measurement. Establish long–short, tall–short
38	Structured apparatus. Arrange big–little, large–small
39	Drawing. Draw straight–curved, straight–crooked lines
40	Solid shapes. Arrange in big–little, large–small sets
41	Explore cutting clay into halves
42	Paper folding. Fold to make: big–little, large–small shapes
43	Plane shapes. Arrange in sets: big–little

Stage test

Stage 4

Number	Activity
75	Time. Explore hour-glass or egg-timer
76	Time. Make–explore sand-clock
77	
78	Counting frame. Use counting-frame to add
79	Shopping. Shopping – one item
80	Weight. Use simple balance
81	Volume. Compare metric measures
82	Number systems. Draw hands to represent numbers
83	
84	Sets. Sort into: tall–short, wide–narrow, thick–thin
85	Geoboard. Make wide–narrow, thick–thin shapes
86	Simulation gaming. Number jigsaws
87	Linear measurement. Arbitrary measures, establish: wide–narrow, thick–thin
88	Structured apparatus. Sort into: more–less, larger–smaller
89	Draw: thick–thin, large–small, wide–narrow lines
90	Solid shapes. Sort into: wide–narrow, thick–thin shapes
91	Explore ½, ¼ kg, 1 lb and balance
92	Paper folding. Fold into: wide–narrow, thick–thin shapes
93	Plane shapes. Sort into: long–short, tall–short shapes

Stage test

The infant mathematics network

This design study is a combined network of the early activities in number work, length volume, weight and time (figure 24, pp. 32–3).

It is envisaged that the pupil will start at the point at which tests show he should begin, and then work down through teaching materials as shown. Using this network either existing teaching materials can be allocated to the activities or teachers can design and produce their own to this pattern.

Length. The experiences start with measuring using arbitrary lengths such as hands, feet, books, etc. After experiences of this type the pupil will refine his measurement and use standard measures. His use of metric units of measurement will be refined until he is able to use his judgement as to which unit to use for the task in hand. For activities 113–118 in the third column it is envisaged that strips will be prepared and colour coded for identification.

Volume. In the columns about volume the same progression from arbitrary units, combinations of units, and then the introduction of metric units used one at a time, is followed by the use of several units to measure volume. One storage problem is the collection of tins, bottles, jars, etc. which are needed to give pupils the requisite variety of experiences of measuring volume. There is no easy way round this as containers do take up a large amount of space. However, some may be stacked and others stored one inside another.

Weight. The lines of activities concerned with weight are arranged in basically the same manner. The first discriminations are the comparisons of objects held in the hands. These objects should be varied in weight and volume. Experience is needed of holding masses which weigh heavily but have little volume, as well as large but light objects. The first stage is to compare two objects widely dissimilar in weight and then further pairs of objects with the difference gradually becoming less obvious. This can be done by having tins or other containers filled with differing amounts of plaster. These can have a letter, number or colour code painted on them. Once the differences between any two objects have been fully explored a differentiation between any three can be attempted.

The next stage will be measuring by means of a comparison with arbitrary units of measurement such as marbles or other objects which have to be counted. The various types of spring scales and balances should be used at the next stage, then metric units of measurement introduced. The first stage of the use of metric units will be using one unit at a time, and later two or more units.

Time. The measurement of time begins with the division of the school day into 'before' and 'after' various events. This is followed by the exploration of time using various methods of measurement. These will include sand-timers, stop-watches and pinger-timers. Comparisons of what can be done during a certain time will be explored. If these networks are used as the basis for the design and production of teaching materials all the stages can be clearly identified and the progression graded as appropriate.

Decimal currency. Before the introduction of decimal currency into Britain it was thought that there would be more problems than in fact there were. A network of activities involving decimal currency was drawn and used as the basis for the design of teaching materials (figure 25, p. 34).

These work cards were modified after use, and additions were made to increase the amount of practice and reinforcement for slower learners. This teaching material has now been incorporated into the total provision of teaching material in use in the school.

Early number activities. The headmaster, was concerned about early number work. He asked if a network analysis could be designed and drawn which involved only those practical activities required for an understanding of number concepts. It was thought that the best approach would be a series of graded exercises using the various types of apparatus in the school. When this network had been drawn, critically scrutinised and modified, it was also used as the basis for the production of suitable teaching materials.

Since all the teachers were experienced infant teachers it was not necessary to specify the exact nature of the activity which would be required at each level (figure 26, p. 35).

A concerted effort by all members of staff enabled them to design and produce an adequate supply of extra teaching material to try to remedy the weakness in number work which had been found earlier. This teaching material also has been incorporated into the total supply of teaching material in the school. Once the material had been incorporated into the work of the school the links between number and weight, number and length, number and volume, and so on, could be drawn. In this way each concept is interlinked and reinforced by all the others.

Interdependences. One of the benefits of using a network of a curriculum is that since it is laid out on one side of one sheet it is easier to see where one aspect of the curriculum is dependent upon another. Then steps can be taken to ensure that the pupil has all the necessary skills before attempting any task.

There is another value in this type of layout when identifying interdependences between subjects. There is a convention that where an activity is dependent upon another this is shown by means of an arrow with a dotted line connecting the two activities. For example in the science network (pp. 49–56), activity number 331 involves a comparison between the sound made by drums of

(1) Sets 1	(51) Sets 2	(101) Sets 3	(151) Sets 4	(201) Sets 5	(251) Sets 6	(301) Dienes 6	(351) Dienes 7
(2) Unifix 1	(52) Unifix 2	(102) Unifix 3	(152) Unifix 4	(202) Unifix 5	(252) Number line 6	(302) Cuisenaire 7	(352) Cuisenaire 8
(3) Cuisenaire 1	(53) Cuisenaire 2	(103) Cuisenaire 3	(153) Cuisenaire 4	(203) Cuisenaire 5	(253) Cuisenaire 6	(303) Abacus 6	(353) Abacus 7
(4) Number line 1	(54) Number line 2	(104) Number line 3	(154) Number line 4	(204) Number line 5	(254) Abacus 5	(304) Number line 7	(354) Number line 8
(5) Counting frame 1	(55) Counting frame 2	(105) Counting frame 3	(155) Counting frame 4	(205) Counting frame 5	(255) Dienes 5	(305) Hand calculator 1	(355) Hand calculator 2
(6) Number recognition	(56) Abacus 1	(106) Abacus 2	(156) Abacus 3	(206) Abacus 4	(256) Number bonds tape 1	(306) Number bonds tape 2	(356) Number bonds tape 3
(7) Number writing	(57) Dienes 1	(107) Dienes 2	(157) Dienes 3	(207) Dienes 4	(257) Practical subtraction: 0–100 number carrying	(307) Practical multiplication: tens, units, x units, number carrying	(357) Practical division by representation: subtraction 0–9
(8) Counting	(58) Practical addition: 0–20, number carrying	(108) Practical addition: 0–100, number carrying	(158) Number bonds tape recording	(208) Number bonds tape recording	(258) Practical subtraction: carrying units	(308) Practical multiplication: tens, units, x units, number carrying	(358) Practical division by representation subtraction 0–20
(9) Number bonds tape	(59) Practical addition: carrying	(109) Practical addition: carrying	(159) Practical subtraction: 0–9, number carrying	(209) Practical subtraction: 0–20, number carrying	(259) Practical subtraction: 0–100, number carrying	(309) Practical multiplication: hundreds, tens, units, x units, number carrying	(359) Practical division by representation: subtraction 0–100
(10) Addition 0–9	(60) Number bonds tape recording	(110) Number bonds tape recording	(160) Practical subtraction 0–9, number carrying	(210) Practical subtraction 0–20, number carrying	(260) Practical subtraction 0–100, number carrying	(310) Practical multiplication: hundreds, tens, units, x units, number carrying	(360) Revision
Test	Test	Test	Test	Test	Test	Test	Test

LINEAR MEASUREMENT

(12)							
(13) Measure using hands	(63) Measure using hymn book length	(113) Measure using long wood strip	(163) Measure using paces and hands	(213) Arrange wood strips in order	(263) Measure trundle wheel: whole metres	(313) Trundle wheel: measure classroom	(363) Trundle wheel: measure classroom in metres and decimetres
(14) Measure paces	(64) Measure exercise book lengths	(114) Measure long balsa strip	(164) Measure paces and spans	(214) Measure balsa strips	(264) Measure hall	(314) Measure hall	(364) Measure hall
(15) Measure spans	(65) Measure book length	(115) Measure long card strip	(165) Measure paces and fingers	(215) Measure card strips	(265) Measure corridor	(315) Measure corridor	(365) Measure corridor
(16) Measure fingers	(66) Measure hymn book width	(116) Measure short wood strip	(166) Measure spans and fingers	(216) Measure books	(266) Measure playground	(316) Measure playground	(366) Measure playground
(17) Measure reach	(67) Measure exercise book width	(117) Measure short balsa strip	(167) Measure hands and fingers	(217) Measure string	(267) Measure field	(317) Measure field	(367) Measure field
(18) Measure feet	(68) Measure book width	(118) Measure short card strip	(168) Measure reach and fingers	(218) Measure hardboard	(268) Measure outside school	(318) Measure outside school	(368) Measure outside school
Test	Test	Test	Test	Test	Test	Test	Test

Fig. 24. *Another modified network analysis of early mathematical experiences broken down into simple activities so that teachers could design teaching materials in the form of work cards.*

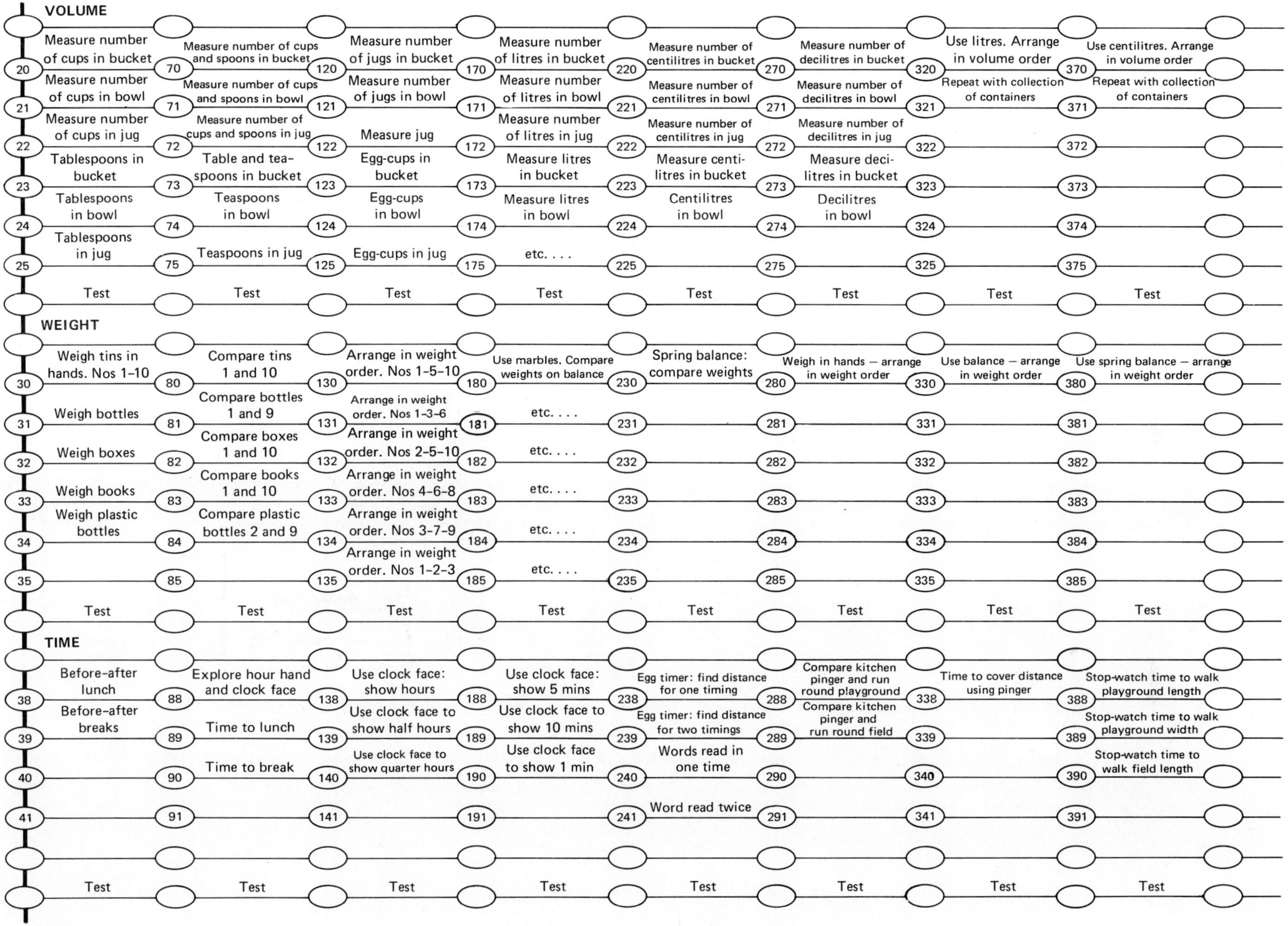

Fig. 24 *(continued).*

VOLUME

Col 1	Col 2	Col 3	Col 4	Col 5	Col 6	Col 7	Col 8
20 Measure number of cups in bucket	70 Measure number of cups and spoons in bucket	120 Measure number of jugs in bucket	170 Measure number of litres in bucket	220 Measure number of centilitres in bucket	270 Measure number of decilitres in bucket	320 Use litres. Arrange in volume order	370 Use centilitres. Arrange in volume order
21 Measure number of cups in bowl	71 Measure number of cups and spoons in bowl	121 Measure number of jugs in bowl	171 Measure number of litres in bowl	221 Measure number of centilitres in bowl	271 Measure number of decilitres in bowl	321 Repeat with collection of containers	371 Repeat with collection of containers
22 Measure number of cups in jug	72 Measure number of cups and spoons in jug	122 Measure jug	172 Measure number of litres in jug	222 Measure number of centilitres in jug	272 Measure number of decilitres in jug	322	372
23 Tablespoons in bucket	73 Table and tea-spoons in bucket	123 Egg-cups in bucket	173 Measure litres in bucket	223 Measure centi-litres in bucket	273 Measure deci-litres in bucket	323	373
24 Tablespoons in bowl	74 Teaspoons in bowl	124 Egg-cups in bowl	174 Measure litres in bowl	224 Centilitres in bowl	274 Decilitres in bowl	324	374
25 Tablespoons in jug	75 Teaspoons in jug	125 Egg-cups in jug	175 etc....	225	275	325	375
Test	Test	Test	Test	Test	Test	Test	Test

WEIGHT

Col 1	Col 2	Col 3	Col 4	Col 5	Col 6	Col 7	Col 8
30 Weigh tins in hands. Nos 1–10	80 Compare tins 1 and 10	130 Arrange in weight order. Nos 1–5–10	180 Use marbles. Compare weights on balance	230 Spring balance: compare weights	280 Weigh in hands — arrange in weight order	330 Use balance — arrange in weight order	380 Use spring balance — arrange in weight order
31 Weigh bottles	81 Compare bottles 1 and 9	131 Arrange in weight order. Nos 1–3–6	181 etc....	231	281	331	381
32 Weigh boxes	82 Compare boxes 1 and 10	132 Arrange in weight order. Nos 2–5–10	182 etc....	232	282	332	382
33 Weigh books	83 Compare books 1 and 10	133 Arrange in weight order. Nos 4–6–8	183 etc....	233	283	333	383
34 Weigh plastic bottles	84 Compare plastic bottles 2 and 9	134 Arrange in weight order. Nos 3–7–9	184 etc....	234	284	334	384
35	85	135 Arrange in weight order. Nos 1–2–3	185 etc....	235	285	335	385
Test	Test	Test	Test	Test	Test	Test	Test

TIME

Col 1	Col 2	Col 3	Col 4	Col 5	Col 6	Col 7	Col 8
38 Before–after lunch	88 Explore hour hand and clock face	138 Use clock face: show hours	188 Use clock face: show 5 mins	238 Egg timer: find distance for one timing	288 Compare kitchen pinger and run round playground	338 Time to cover distance using pinger	388 Stop-watch time to walk playground length
39 Before–after breaks	89 Time to lunch	139 Use clock face to show half hours	189 Use clock face to show 10 mins	239 Egg timer: find distance for two timings	289 Compare kitchen pinger and run round field	339	389 Stop-watch time to walk playground width
40	90 Time to break	140 Use clock face to show quarter hours	190 Use clock face to show 1 min	240 Words read in one time	290	340	390 Stop-watch time to walk field length
41	91	141	191	241 Word read twice	291	341	391
Test	Test	Test	Test	Test	Test	Test	Test

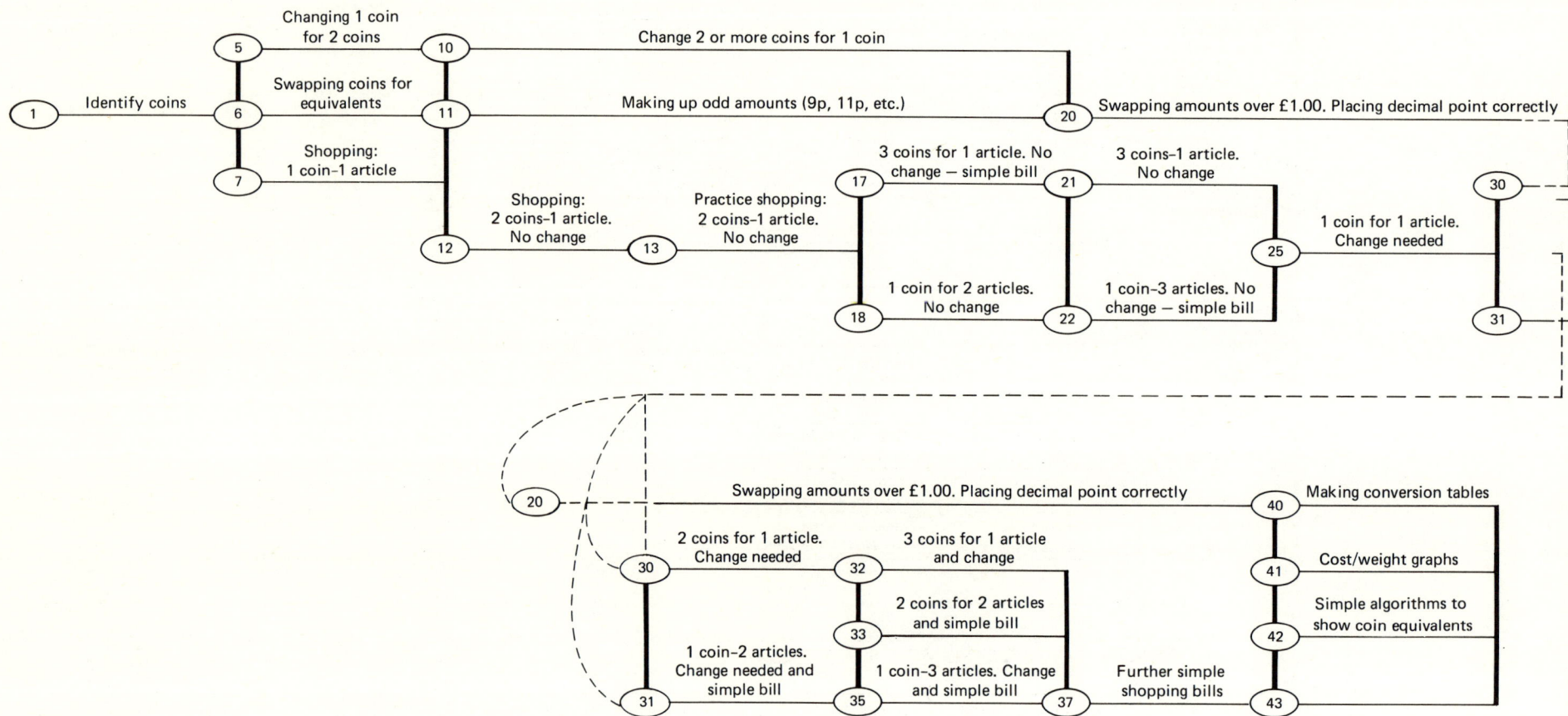

Fig. 25. *Simplified network analysis of activities involving pupils early experiences using decimal currency.*

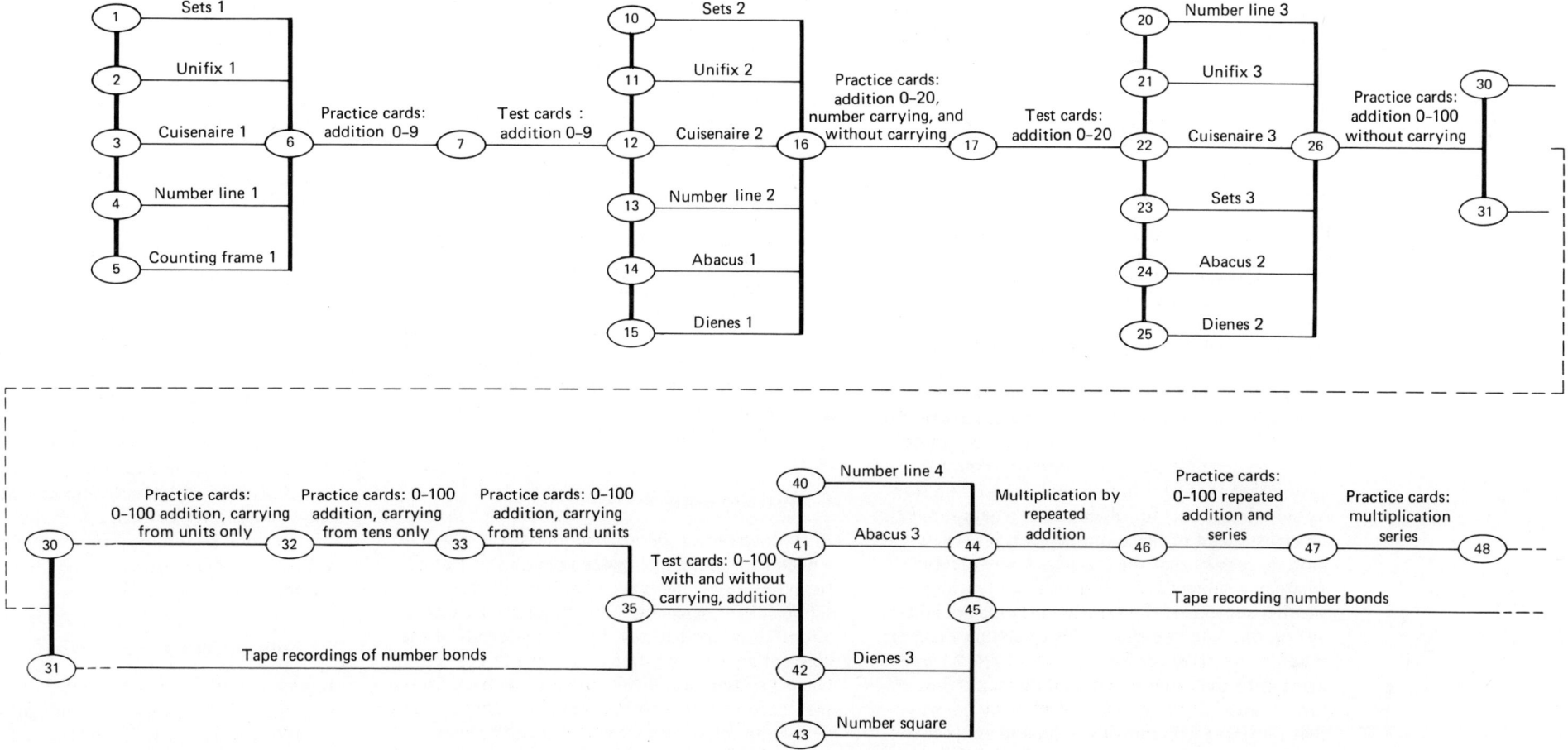

Fig. 26. *Simplified network analysis of early activities in number using various types of structured apparatus.*

varying sizes. While the contrast between the very large and the very small is obvious, at some stage it will be necessary to measure the diameters of drums and compare the sounds. The measurement of diameters of circles is activity number 726 on the mathematics network (figure 22*b*) and the dependency can be shown as in figure 27.

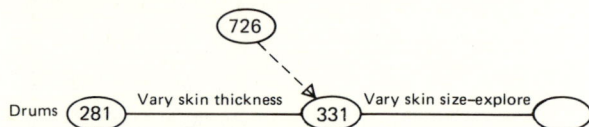

Fig. 27. *The method of showing the dependency of one activity upon another by broken arrows and a number. In this case activity 331 on the science network (figure 39) is shown to be dependent upon activity 726 of the mathematics network figure 22b.*

In the primary school usually one teacher is responsible for most of the teaching in a particular class. If the attention of the class teacher can be drawn to dependencies, steps can be taken to ensure that the work in two subjects goes on side by side. In schools for older pupils where subject specialists tend to teach one subject, it is even more necessary that the teachers know precisely where one subject is dependent upon teaching in another subject. If the various curricula are laid out on one side of a sheet, like a network diagram, it will be much easier for teachers to reach agreement as to the order in which they tackle the various areas of their own subject in such a way that they take account of teaching in another subject. If the dependencies are clearly shown by dotted arrows, or in some equally simple manner, the teacher can check that the pupils have mastered the necessary skills required, before going forward.

Implementation. Once such a syllabus has been devised, it is necessary to have a full discussion with all the staff involved to be sure that every member knows exactly what his part in the overall scheme is and how it is to be carried out. The agenda of a staff meeting of this kind would be:

1   An explanation of the modified technique of network analysis.
2   The outline of a scheme for the use of the network analysis of the school mathematics syllabus as a basis for teaching mathematics throughout the school.
3   The allocation of the various responsibilities to members of the staff. (Figures 20, 21.)

Among the duties of staff members would be:

1   Assembling and numbering teaching materials.
2   The design and implementation of a simple storage and retrieval system
3   The allocation of specific parts of the school for different purposes.
4   Training the pupils to be self-reliant so that they can obtain the materials and apparatus as they need them, and use the storage and retrieval system sensibly.

1   Teaching materials

In an unstreamed school in any class there will be a need for a variety of teaching materials suitable for pupils of a wide range of ability. In many schools for younger children materials are either allocated on the basis of the average needs of the class, or are in a common store for the year group. Once a system of individual work is started, pupils may need materials from other class stores or other year stores. Any teacher who has had to seek materials which are allocated on this basis will know how inconvenient this can be, and how frustrating for the pupil.

With the great variety of teaching materials available today careful thought and planning must go into deciding which items should be in a central store and which in a class area. The larger items or those which are not in constant demand might be allocated to a central store, for example films, film strips, slides, tape recordings, video-tape recordings and records. Items which might be allocated to the class area are compasses, protractors, stop-watches, clocks, scales, mathematical balances, etc. In the class area there ought also to be a comprehensive series of work cards and graded text books. These would be numbered according to the numbers given to each activity in the network analysis.

When the process of collecting and allocating the school stock is in hand it is a good time to consider what further items might be required. Since the network will set out the activities undertaken by the pupils in a step-by-step fashion, it is fairly easy to work through and decide which types of apparatus and which materials will be required. Any shortfall in stock can then be made good.

Although it might not seem so, this is a more economical way of distributing apparatus than in class units. When each class needs a set of items, one for each pupil, the total required can be very large indeed. In allocating apparatus as required by individual pupils, such quantities are no longer needed. Each pupil will only need for example, one protractor at intervals; there is no need for a class set of thirty or forty. Although a diversity of items will be required for an integrated individualised scheme there will be no need for large numbers of any one item.

2   Storage and retrieval

In an integrated scheme in which pupils will be working as individuals some thought must be given to the design and construction of storage units so that materials are readily available. Storage will be needed for teaching materials, record cards, and for apparatus and materials.

Teaching materials store. The largest collection of teaching material in a school will probably be in the form of work cards. The largest of these will probably be A4 in size. Thus all will fit into box files designed for A4 sized sheets (21 cm x 30 cm).

A box 18 cm x 25 cm x 30 cm has been found to work well. These boxes can be stored on shelves or in cupboards. Since they will be in constant use they should be in a place which is easy for pupils to reach, and not likely to be congested. If the materials used take up more than one box it might be better to place them in different parts of a room to spread the crowd of pupils who may gather round. The box files can be made from scrap-wood, and if parents can be persuaded of the value of the scheme some are sure to give willing help. The design in figure 28 has proved successful using two pieces of blockboard and three pieces of hardboard, glued and panel pinned. The dimensions of the pieces are: blockboard: 2 x (24 x 18) cm and hardboard: 3 x (30 x 15) cm. Empty boxes can easily be stacked. If extra support is required for the teaching materials the back of the files can be made taller.

Record card stores. Pupils' record cards (see p. 42) are best kept in box files in alphabetical order. Some teachers find it easier to handle record cards if they are stored in ring files. If tabs are used for indexing, they should be of the type which can be moved. There are sure to be new pupils arriving and others departing during the school year.

If the record cards are to be stored in a box file they should either be printed or mounted on card to enable them to stand upright without bending. If a ring-binder is used the cards can be printed on paper or thin card and provided they have gummed strengtheners stuck round the ring hole they will last well.

Audio-visual teaching material store. It is convenient to have the main card index in the library, listing all the work materials which are in book form. There also needs to be a card index for films, film strips and loops, tapes and video-tapes. Each will have its own card index and this should be kept near the appropriate storage area.

Films, tape recordings, and video-tape record-

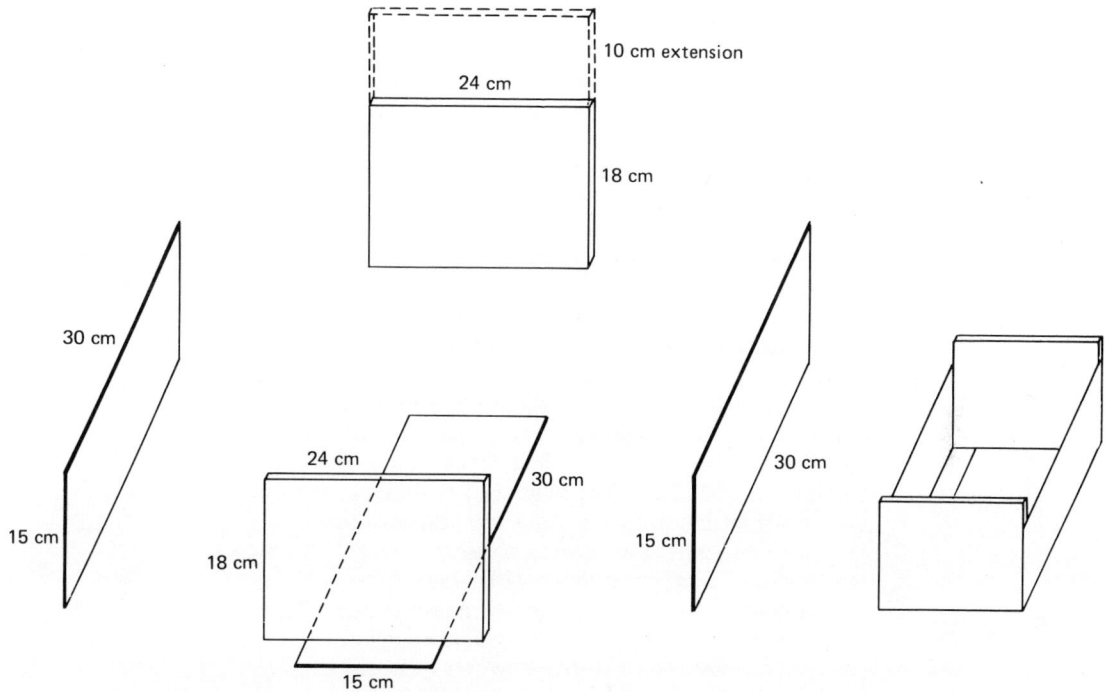

Fig. 28. *Exploded diagram of simple box file for teaching materials used in integrated mathematics scheme.*

ings can be conveniently stored on shelves on edge. For quick easy checking strips of paint should be painted on diagonally (figure 29). Since work cards, record cards and index cards are going to be heavily used, some thought should be given to preserving them. Cover them with plastic bags or film. There are also methods of coating card thermally, or they can be given a coat of varnish.

Slides are most easily stored in magazines. Overhead projector transparencies can be stored in ring-binders with interleaved cards. These can be stored on shelves with a card index nearby. A diagonal strip of paint will assist in the easy checking of these also.

Film strips can be stored in shallow boxes. Dividers can be made from strips of card cut halfway and interlocked (figure 30).

At the same time decisions will have to be made about the placing of storage shelves, trays, indexes and drawers. The visual aids should be kept in an area which can be darkened, and which has white walls or permanent screens. Since many visual aids are heavy, and in any case ought not to be shaken when the elements are hot, they are best used in one area and moved as little as possible.

Audio aids are rather more easily moved, especially now that transistors are commonly used in their construction and they are lighter, more compact, and sturdier. It may however be of value to have the heavier and more bulky apparatus gathered at one point, with the teaching materials and index to hand. Even with more sturdy construction, generally the less such equipment is moved the less likely it is to be damaged.

To avoid interference, full use should be made of headphones with audio aids. These may be plugged into a junction box and can be easily controlled by a child. Headphones with a built-in miniature receiver can enable operations to be carried out freely by the child while receiving instructions from suitable equipment.

It is fairly simple to make carrels from sheets of hardboard cut half-way as in the diagram in figure

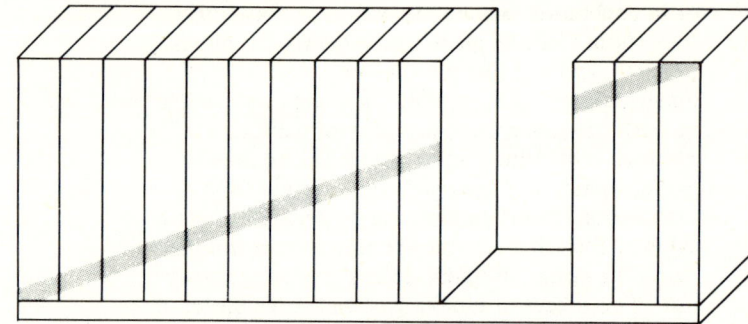

Fig. 29. *Method of storing audio or video tapes on a shelf with a diagonal stripe to enable an easy visual check of missing items.*

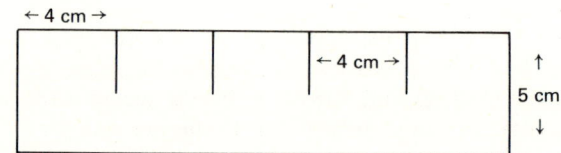

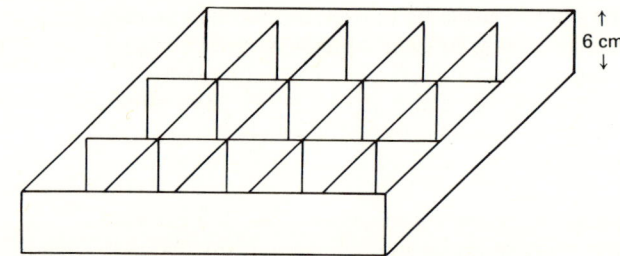

Fig. 30. *A simple method of storing film strips.*

31. These can be used for the audio aids or teaching machines and prevent the pupils being bothered by the actions of others. These naturally need to be sited conveniently near the materials and programmes.

3 Allocation of furniture and space

It is clear that some thought will need to be given to the moving of furniture, and other additions to existing stock. Much of what is needed can be adapted or modified. It is probable that extra shelving will be needed but that is not difficult to arrange.

Space will also have to be allocated for particular purposes. If activities involving experiments with water are to be included in the scheme, a 'wet' area needs to be designated and equipped. A 'noisy' area will also be needed for activities such as listening in groups to broadcasts and recordings and using musical instruments. The positioning of this area will affect the choice of the 'quiet' area which will also be needed. The placing of audio-visual aids and teaching materials of all kinds, including reference books in a library area must also be carefully planned.

4 Pupil training

When a class worked as a unit, or in small groups, it was possible for the teacher to act as a dispenser of materials and equipment, as well as being a source of information. Once an attempt is made to enable each child to follow the course bested suited to his needs, this rôle becomes at the same time less necessary and also almost impossible. The school resources in the form of teaching materials take the place, to some extent, of the teacher as a source of information. This implies that the child is capable of finding the materials which are needed, using them in a sensible manner, and can then return them to the correct place in the store. The teacher cannot be expected to find all the materials

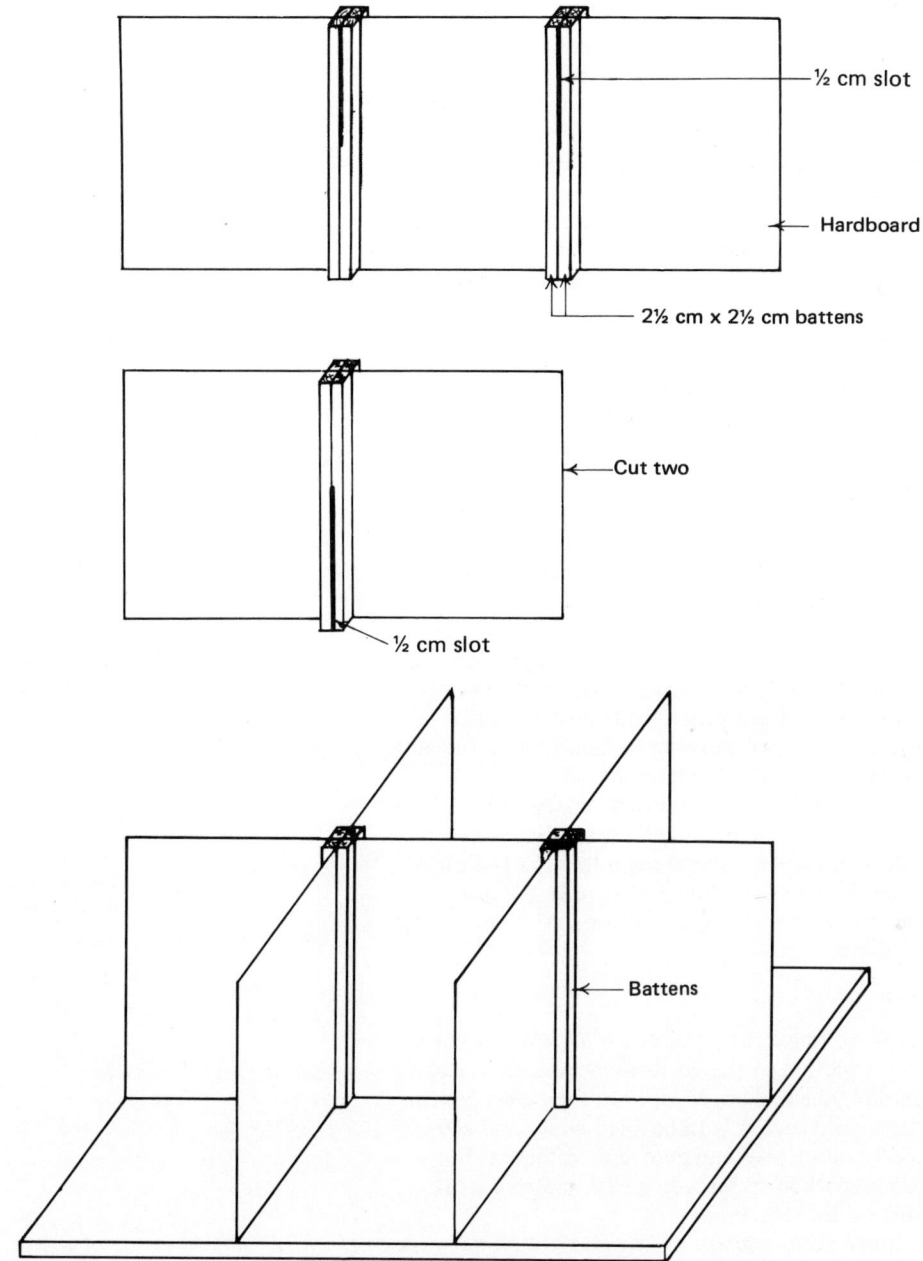

Fig. 31. *Diagram to show construction of simple carrels for use with audio aids or teaching machines.*

which will be required by a class of active children, nor will he be able to give individual instruction in the use of the materials, and ensure that they are all put away properly. There must be a deliberate policy of retraining the pupils to take advantage of the freedom which is presented by individual working, but this freedom must be under overall control. The pupils must be deliberately trained. This entails a new rôle for the teacher. He has to stand back and tell the child how to find what he needs, instead of supplying the need himself.

The network will need to be explained to the children in simple terms, but experience has shown that there need be no great problem here. If the children are started off in groups, feedback can be used to modify the approach if it is found that there are snags. If the children are accustomed to the use of simple teaching algorithms the examples in figures 32 and 33 can be used for this purpose.

The pupils will also have to be taught to use an index, how to find the various resources in the school, and how to return the resources sensibly. Ancillary staff can play a valuable part in this process, but they too must be trained to help the child find what he needs for himself.

If this training is to take full advantage of the opportunity to work at their individual pace, the children must be given enough help, but not too much. The decision how far to go is a delicate one which must be left to the judgement of the teacher.

Pupil records

It will no longer be possible for a class teacher to keep a picture of the position of each child clearly in mind when all are engaged in their own course. If the child's work is to be adequately and efficiently monitored and guided an effective, but simple method of recording each piece of work must be devised.

Since every activity on the curriculum network has a unique number all that is required is a list of

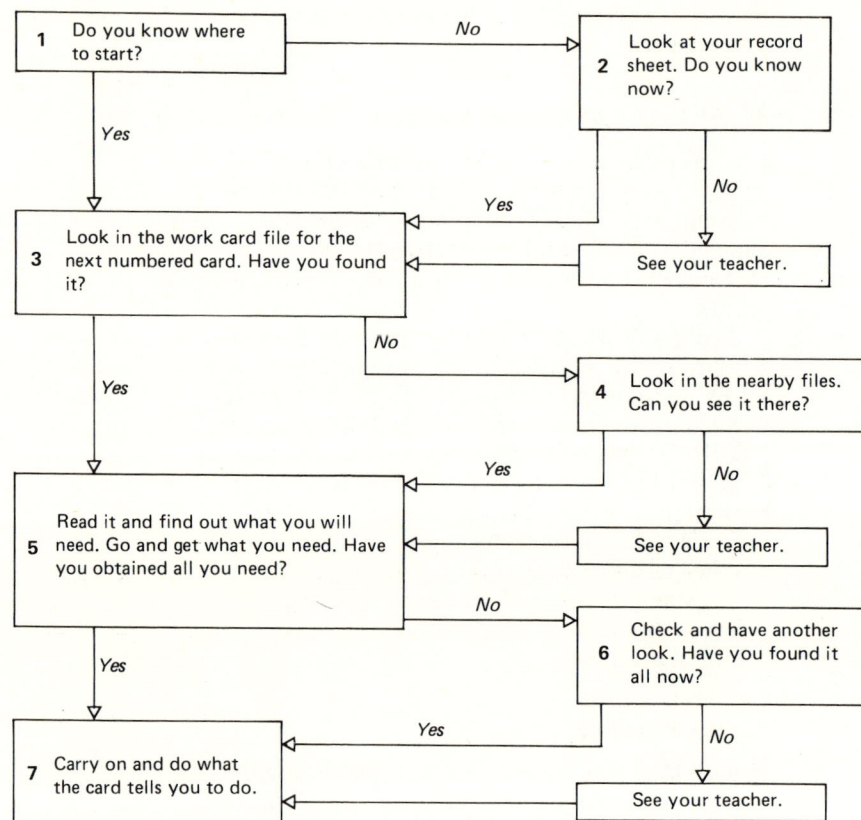

Fig. 32. *Simple algorithm to illustrate activities of pupil using individualised scheme of work.*

numbers on a card which corresponds to the numbering system. A final version of a record card, the result of testing and modification, is shown in figure 34. It is simple, easy to use and effectively and clearly shows what the pupil had done, how many times and at what level of difficulty. The activity number is printed beside nine small squares. These show the three levels of difficulty of the teaching materials. The spaces allow for three attempts at each level of difficulty if necessary (figure 35, p. 43).

The lines of activities on the network of the curriculum are copied on the record card. For example activities: 1, 25, 50, 75, 100, etc. follow a continuous line of activities on the network. This was a particular help in diagnosing areas where the pupils were experiencing difficulties. A simple system was used for making entries. The small section of a record card in figure 36 shows the system in use.

The pupil to whom the record card in figure 36 relates has completed activities 103, 104 and 105 in stage 5 successfully at the most difficult level. This would probably be a pupil of considerable ability. He has continued successfully completing activity 130, but failed to complete activity 131 correctly and so completed the same activity at a moderately difficult level with success. He then tried again at a difficult level with success. This type of record card shows a profile of each pupil and can be checked at a glance. The pupil proceeds down a column of activities numbered to correspond with the network. These columns are termed stages. When he has completed a stage he takes a test of the skills which he should have mastered. The result of the stage test is entered in the space at the foot of that column.

The result of the test will enable the teacher to decide whether the pupils should undertake that series of activities again at another level of difficulty, or carry on to the next column of activities. If the pupil has failed to master some particular

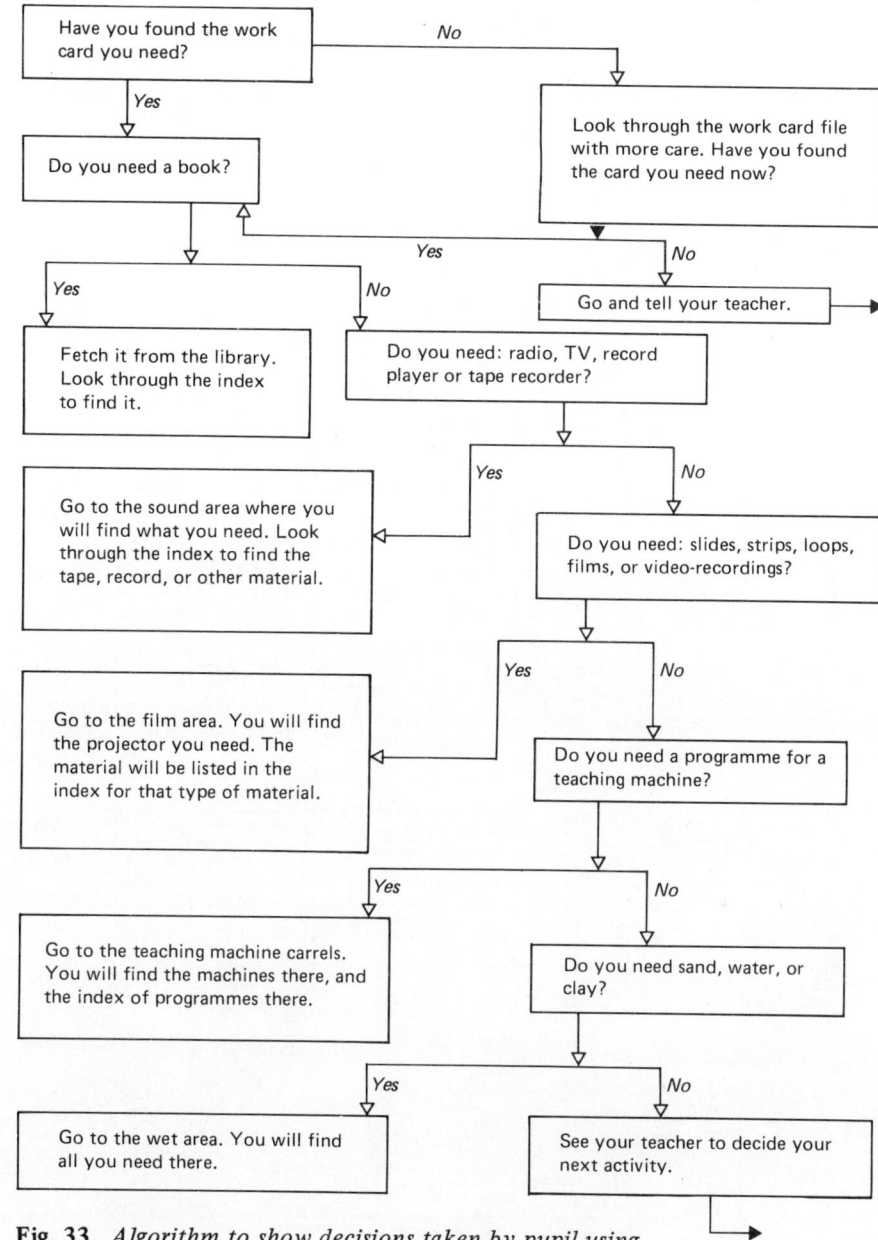

Fig. 33. *Algorithm to show decisions taken by pupil using different teaching materials in integrated scheme.*

PUPIL RECORD CARD   NAME :-

B.W. Vaughan © 1973

Mathematics

	Stage 1	Stage 2	Stage 3	Stage 4	Stage 5	Stage 6	Stage 7	Stage 8	Stage 9	Stage 10	Stage 11	Stage 12	Stage 13	Stage 14	Stage 15	Stage 16	Stage 17	Stage 18	Stage 19	Stage 20	Stage 21	Stage 22	Stage 23	Stage 24	Stage 25	Stage 26	Stage 27	Stage 28	Stage 29	Stage 30
1						130	160	190	225	260	300	340	380	420	463	500	540	580	620	660	700	740	780	820	860	900	940	980	1020	
2	1	25	50	75	100	131	161	191	226	261	301	341	381	421	464	501	541	581	621	661	701	741	781	821	861	901	941	981	1021	
3															465	502	542	582	622	662	702	742	782	822	862	902	942	982	1022	
4	2	26	51	76	101	132	162	192	227	262	302	342	382	422	466	503	543	583	623	663	703	743	783	823	863	903	943	983	1023	
5														423	467	504	544	584	624	664	704	744	784	824	864	904	944	984	1024	
6	3	27	52	77	102	133	163	193	228	263	303	343	383	424	468	505	545	585	625	665	705	745	785	825	865	905	945	985	1025	
7					103	134	164	194	229	264	304	344	384	425	469	506	546	586	626	666	706	746	786	826	866	906	946	986	1026	
8	4	28	53	78	104	135	165	195	230	265	305	345	385	426	470	507	547	587	627	667	707	747	787	827	867	907	947	987	1027	
9					105	136	166	196	231	266	306	346	386	427	471	508	548	588	628	668	708	748	788	828	868	908	948	988	1028	
10					106	137	167	197	232	267	307	347	387	428	472	509	549	589	629	669	709	749	789	829	869	909	949	989	1029	
11	5	29	54	79	107	138	168	198	233	268	308	348	388	429	473	510	550	590	630	670	710	750	790	830	870	910	950	990	1030	
12	6	30	55	80	108	139	169	199	234	269	309	349	389	430	474	511	551	591	631	671	711	751	791	831	871	911	951	991	1031	
13					109	140	170	200	235	270	310	350	390	431	475	512	552	592	632	672	712	752	792	832	872	912	952	992	1032	
14	7	31	56	81	110	141	171	201	236	271	311	351	391	432	476	513	553	593	633	673	713	753	793	833	873	913	953	993	1033	
15	8	32	57	82	111	142	172	202	237	272	312	352	392	433	477	514	554	594	634	674	714	754	794	834	874	914	954	994	1034	
16	9	33	58	83	112	143	173	203	238	273	313	353	393	434	478	515	555	595	635	675	715	755	795	835	875	915	955	995	1035	
17	10	34	59	84	113	144	174	204	239	274	314	354	394	435	479	516	556	596	636	676	716	756	796	836	876	916	956	996	1036	
18					114	145	175	205	240	275	315	355	395	436	480	517	557	597	637	677	717	757	797	837	877	917	957	997	1037	
19	11	35	60	85	115	146	176	206	241	276	316	356	396	437	481	518	558	598	638	678	718	758	798	838	878	918	958	998	1038	
20	12	36	61	86	116	147	177	207	242	277	317	357	397	438	482	519	559	599	639	679	719	759	799	839	879	919	959	999	1039	
21	13	37	62	87	117	148	178	208	243	278	318	358	398	439	483	520	560	600	640	680	720	760	800	840	880	920	960	1000	1040	
22	14	38	63	88	118	149	179	209	244	279	319	359	399	440	484	521	561	601	641	681	721	761	801	841	881	921	961	1001	1041	
23	15	39	64	89	119	150	180	210	245	280	320	360	400	441	485	522	562	602	642	682	722	762	802	842	882	922	962	1002	1042	
24	16	40	65	90	120	151	181	211	246	281	321	361	401	442	486	523	563	603	643	683	723	763	803	843	883	923	963	1003	1043	
25	17	41	66	91	121	152	182	212	247	282	322	362	402	443	487	524	564	604	644	684	724	764	804	844	884	924	964	1004	1044	
26					122	153	183	213	248	283	323	363	403	444	488	525	565	605	645	685	725	765	805	845	885	925	965	1005	1045	
27	18	42	67	92												526	566	606	646	686	726	766	806	846	886	926	966	1006	1046	
28					123	154	184	214	249	284	324	364	404	445	489	527	567	607	647	687	727	767	807	847	887	927	967	1007	1047	
29																		608	648	688	728	768	808	848	888	928	968	1008	1048	
30								215	250	285	325	365	405	446	490	528	568	609	649	689	729	769	809	849	889	929	969	1009	1049	
31									251	286	326	366	406	447	491	529	569	610	650	690	730	770	810	850	890	930	970	1010	1050	
32									252	287	327	367	407	448	492	530	570	611	651	691	731	771	811	851	891	931	971	1011	1051	
33												368	408	449	493	531	571	612	652	692	732	772	812	852	892	932	972	1012	1052	
34																				693	733	773	813	853	893	933	973	1013	1053	
35	19	43	68	93	124	155	185	216	253	288	328	369	409	450	494	532	572	613	653	694	734	774	814	854	894	934	974	1014	1054	
36									254	289	329	370	410	451	495	533	573	614	654	695	735	775	815	855	895	935	975	1015	1055	
37									255	290	330	371	411	452	496	534	574	615	655	696	736	776	816	856	896	936	976	1016	1056	
TESTS 1																														
TESTS 2																														
TESTS 3																														

Fig. 34. *Pupil record card based on the modified network analysis for pupils ages 5–13 years (figure 22b, parts 1–6, pp. 24–9).*

activity the teacher may decide that he should enter a remedial sequence of activities. Among the teaching materials will be ranges of cards in order of difficulty to enable the child to carry out a series of related activities aimed at gaining mastery of a particular aspect of mathematics.

The term 'work cards' is used as a matter of convenience and must be taken to refer to any form of teaching material whether printed or audio-visual. Cards of a standard size are easy to store and there are many published printed work cards which can be numbered according to the network numbering system. In addition cards can be inserted to tell a pupil to take a book, turn to a page and carry out an exercise, to take a tape recording or cassette, and carry out the instructions recorded or to use slides or film strips. In this way any available teaching materials can be integrated into the course.

As the record card is completed the progress of each individual pupil is presented as a profile, showing exactly where he has made progress, and at which activities he has had difficulty. It will also be apparent where thought needs to be given to supplementing teaching materials, and providing remedial materials. The records will also show clearly which pupils are not making the progress they ought to make, and those who are working to capacity. The child who has mastered an activity at the first attempt can go on to the next item of teaching material. The pupil who needs reinforcement or practice will take another item giving further experiences. To sum up: the record cards will show the progress of each pupil at a glance, and at which stage he has had problems.

Such a scheme provides an infinitely flexible system, which enables the pupil to proceed rapidly when he can gain mastery of the material, but provides extra teaching material when he has difficulty. It enables teachers to design teaching materials where the pupil records show that they are most needed. It is also of value in the design of remedial material, and should enable the causes of

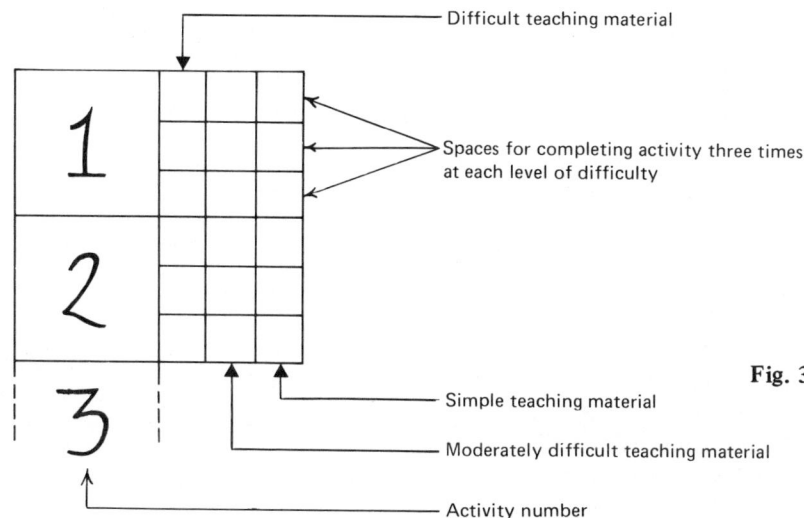

Fig. 35. *Two activities from pupil record card showing spaces for entries to indicate level of difficulty of task completed and number of attempts at the activity.*

Fig. 36. *A section of pupil record card showing entries made by teacher. In stage 5 activities at difficult level completed successfully. In stage 6, activity completed at difficult level (130) and two at moderately difficult level (131).*

pupil failure to be identified, by tracing back through the network.

When each record is passed from teacher to teacher, the exact position of the pupil in the overall scheme can be pinpointed, and each pupil can carry on without any break. This should render unnecessary that difficult period when a child moves to a new class, when the teacher has to decide which stage he has reached. The records should also be of value if passed from school to school. If the receiving school has a copy of the network, and record card, the position of the pupil can be accurately placed, and future work geared to his needs. If the school is not using the same syllabus the record card will enable an accurate objective assessment to be made of the pupil's work. Attainment, speed of progress and ability can be described exactly and vague and subjective judgements culled from memory will be avoided.

The school record card

This would not need the same detail as the class record card. It could have a format similar to figure 37.

The class teacher might send a record of completed activities to the school secretary once each week to make up a central pupil record. This would also contain the results of any tests. Physical, or psychological factors which might affect progress could also be entered on the central record card.

The results of the stage tests only might be sufficient. A line could be ruled down a column to indicate completion of a stage, and the result of the stage test entered at the foot of the column. If the pupil was guided to take the same column of activities again at a different level, a second line ruled down the column would show this, together with the second score in the stage test.

The number of lines ruled down the columns would soon show which columns were causing the most difficulty and help to identify the cause. School policy would decide how often activities

NAME: ______________________ Date of birth __________

ADDRESS: ______________________

DIAGNOSTIC TESTS: ______________________

ATTAINMENT TESTS: ______________________

STAGES														
1	2	3	4	5	6	7	8	9	10	11	12	13	14	15
STAGE TEST SCORES														

Note:
There may be a space for additional comment.
Punched holes at the top may be useful for quick sorting.

Fig. 37. *Specimen school record card.*

should be repeated before passing on to a new activity. With the proper use of such records the amount of routine clerical work would be kept to a minimum, while giving maximum value in recording pupil progress. They should also be of benefit in directing the attention of teachers to areas in which most effort is needed.

Work cards

Cards should be of standard size to facilitate storage. Once they are printed, a surface should be added to preserve them from damage. This might take the form of:

1 A simple plastic bag.
2 A coating of polish; this serves for a time.
3 Painting with a matt floor sealer or varnish. This lasts well but takes time to apply. This is an activity in which parents might be involved, if they can be persuaded of the value of the exercise.
4 A plastic coating heat-sealed on to the cards. This is probably the best solution but expensive.
5 The application of an adhesive acetate sheet. Also effective but expensive.

It is worth taking some trouble over this operation. Once cards become illegible they produce a constant series of queries which are time-consuming to answer and unproductive. Scruffy cards will give children a poor view of the work, and conversely good work can be expected when cards are neat and attractive in appearance.

The need to train the children to use the storage and retrieval system properly has already been mentioned. Experience shows that no matter how well trained the children are, there are always some who do not do as they are shown for one reason or another.

In practice pupils have little difficulty in finding their work cards and other teaching materials, but those who can find cards are not so careful about returning them. It is a good policy to instruct the pupils to put used cards in a pile in a convenient place. At a suitable time one reliable pupil can replace them all. This works very well. At intervals the teacher, ancillary, or another pupil can check through the whole set of cards to ensure that they are in the correct order.

The introduction into the class: day-to-day running

The teacher will have a copy of the network and will be familiar with its layout.

He will also have a box file of basic work cards and other teaching materials numbered in accordance with the network. He will also have a record card for each child.

For a period of several weeks he will have taught the children to be more self-reliant. They may have been introduced to the use of algorithms, and may even have tried to make some for themselves. They are likely to enjoy doing this and soon learn the strategies involved. They will also have been taught to use an index and reference books. The storage of all items and materials which the children need will have been explained. Such things as, paper, scissors, ink and paint should not entail a question being asked of the teacher. The children should be encouraged to take what they need, but with care and without extravagance.

Non-readers or poor readers can have their cards read to them; the instructions may be on tape or read by an older child or the teacher.

The children should be introduced to the idea of marking their own work when this is possible. They are capable of doing this honestly for the most part. The risk is in my experience minimal, especially when the children realise that they are not competing with any other child, but are pursuing a course of their own. Nevertheless marking by the child should be taught, and the stupidity of false marking shown.

The network should be explained to the pupils. They will be most interested to see where what they are doing fits into a general scheme, and where it is all leading.

The pupils will then be tested and allocated to an activity in the network which seems most suitable to their individual needs. This should be done a few pupils at a time. This will enable the teacher to gain experience of the method of working, and to make any slight modifications as may seem necessary, without being overwhelmed. The process is set out on the flow chart in figure 38.

It can be summed up as follows:

1 The pupil is allocated an activity.
2 He collects the work card or teaching material.
3 He carries out the activity as directed on the card.
4 He returns the materials used.
5 He takes the completed work to the teacher.
6 The teacher checks that it has been marked correctly.
7 On the basis of the work completed the teacher will decide whether the pupil should:
   a proceed to the next activity in the column, or
   b enter a remedial sequence, if mastery is obviously not likely to be reached, or
   c engage in the same activity at a more advanced level.
8 The record of the activity completed, is entered on the pupil's record card.
9 The pupil is directed to his next activity.

On completion of a column of activities, the pupil will have a stage test administered. This ought to be supervised, and marked by the teacher. It is on the evidence gained by this test that the next stage in which the pupil is to engage, is decided. The evidence gained from tests may also be used to modify the network, or the work based on it.

Work cards can be written at three levels of difficulty. They can easily be distinguished by, for example, cutting off one or other of the top corners. The pupils are directed to take cards according to the teacher's assessment of their ability. This might appear to be inflexible but is not. At any time the teacher can vary the level of difficulty of the material for an individual pupil. Pupils who may be unable to carry out activities of the simplest nature in some areas may be adept at others. The three grades of difficulty of activities enables the teacher to move pupils to a level to meet individual needs at any time. This takes account not only of varying abilities, but also variations from time to time in a pupil's rate of progress.

It did not seem that pupils gained a poor self image if they were directed to undertake work at a simple level, as they were usually only too well aware of their shortcomings. In most classes all the pupils are aware of the rough order of ability and we deceive ourselves if we think we can disguise this. What is important is that the pupils should be given activities geared to their individual needs and at which they have a good chance of success. Once pupils are working on an individual course and are no longer being compared with their peers, so long as they succeed at their individual activity, they are highly motivated to continue. We found pupils enjoyed the scheme and many were disappointed if they had to miss mathematics.

The reason why there are only three levels of difficulty in the scheme described is that it was thought that this would suffice. In practice it was found to work quite well. There is no reason why the number of gradings should not be increased but only a careful study of results will show what the optimum is. Certainly in a remedial sequence there would be a need for much smaller steps, and many more grades of difficulty.

Most of the teaching material in use is in the form of work cards because these are readily available and can most easily be deployed. We also use pages of text books cut up, backed with card, inserted in plastic bags and numbered. Some teaching programmes are also used where they fit into the overall scheme. These are useful when a pupil needs a step-by-step explanation of an essential

operation. They are also of value as remedial sequences, or for pupils who have been absent, to enable them to catch up. We envisage that other teaching materials will be incorporated as the scheme develops.

In the emphasis on individual work the value of group and class work must not be overlooked. We find that each class teacher taking the basic scheme gradually reaches a balance of class, group and individual working. This we think is the right approach. Pupils who are continually working on

their own like it for a time, but there is value in co-operative working with others in a group. There is also value in class work under the close direction of the teacher. The exact division of time between these activities is best left to the judgement of the individual teacher. It is easy for any class teacher given a method of organising and controlling an individualised scheme of work, to introduce a balance of class and group work, as experienced teachers are thoroughly familiar with these techniques.

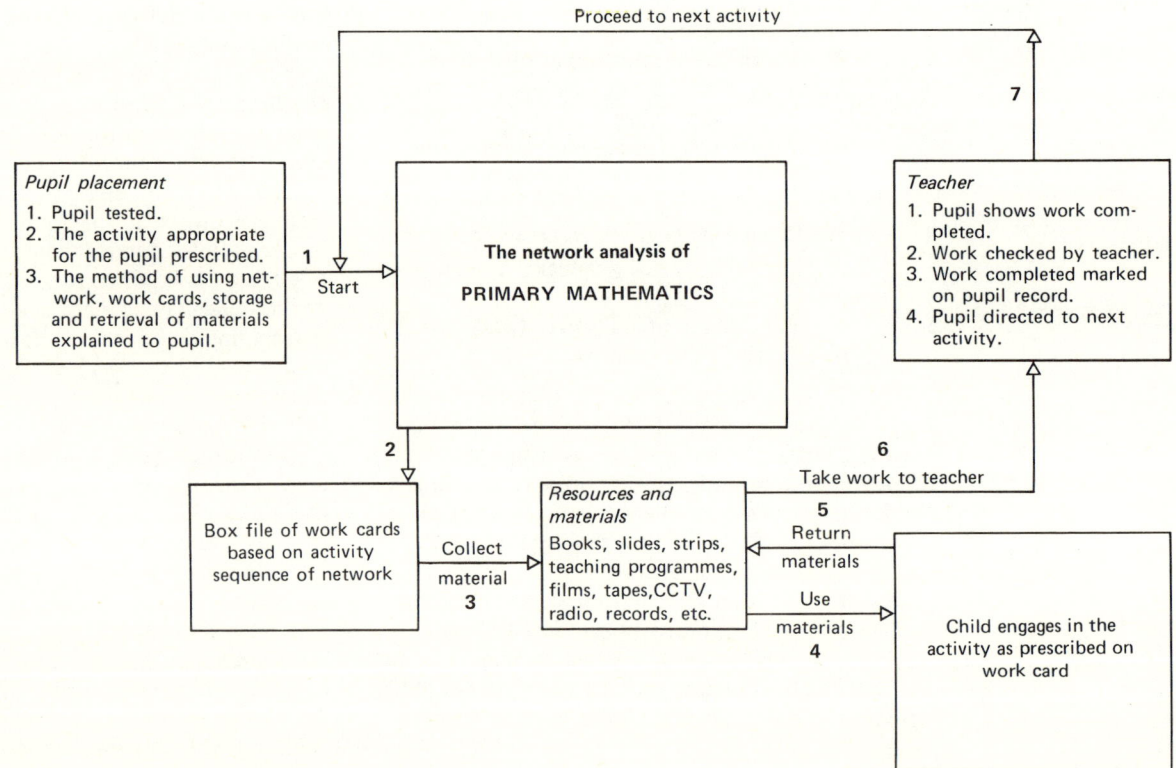

Fig. 38. *A diagrammatic flow chart of the system of using an individualised scheme based upon a simplified network analysis of mathematics.*

5 · Science, art and craft

Science

The changes which have affected schools in recent years all affect the teaching of science. In addition there has been a continual increase in the importance of scientific technology in our lives and therefore of the importance of this subject in our schools. All the reasons for a new look at the layout of the syllabus, and the building up of an integrated scheme which relate to mathematics apply with the same force in the teaching of science. Since science is a subject which has the same sort of logical structure as mathematics, it ought to be susceptible to the same type of treatment with the same advantages.

This account of the manner in which an integrated scheme could be built up in primary science makes use of the experience gained in the development of the mathematics scheme described in the previous chapter. The diagrams setting out a plan for the introduction and the organisation of the scheme for mathematics can be used in a similar scheme for science. The way in which a network analysis may be used to examine particular areas of mathematics in detail so that teaching materials could be designed and produced should be borne in mind.

One of the benefits of any sort of systematic analysis of a subject area is the insights one gains into the dependence of one aspect of the subject upon another, and the interdependences. Even if the network included in this design study is used only as a source of ideas, or simply as an example of the technique it should have value. If the staff of a school took the network and decided to make a selection, concentrating on certain activities it would at least help them to clarify their own thoughts. Ideally each group of teachers should use the technique, to draw up their own curriculum, or else take the design study and modify it as seems best.

The first problem is to decide on a suitable starting point. Since young children are fascinated by their own bodies it seemed that this might provide a series of separate starting points. Thus the start of each line of activities in the upper part of the network is a part of the body of the individual pupil. From this starting point the pupil is led to compare and contrast his own body with that of other living creatures. He is helped to build up ideas about the similarities and differences between animals and plants by making collections, either of objects or of drawings and pictures. The pupil is led to experiment in simple ways to find out more about the world around him.

The lower section of the network is concerned with plants and the same process of discovering similarities and differences takes place.

In the network an attempt has been made to sort out the experiences of the pupils into the same stages of learning as outlined previously. This process of sorting into hierarchies is very similar to that attempted in regard to mathematics.

The pupil starts with a series of experiences to discover gross differences, through his observations and collections of objects, visits to museums, use of visual aids, etc. These experiences enable him to refine his discriminations. From these experiences he is able to abstract the underlying concepts.

The progression from the simple to more complex is the same as for mathematics and during the latter years in a primary school the pupils may be able to see cause and effect, and to draw conclusions from their various experiments.

A similar plan to that used for the mathematics scheme could be used for the introduction of an integrated scheme for science.

This introduction will involve:

> The training and testing of pupils to assess their correct starting point in the scheme.
> The collection and deployment of teaching materials and apparatus.
> Designing and implementing a storage and retrieval system.
> Designing and producing a record card based on the scheme.

The record card would be the same as that produced for the mathematics scheme (p. 42) but the numbering would correspond to that on the science flow chart. It would be used in exactly the same way though it is worth adding a more flexible system of grading for the more extended and open ended activities.

When the design study of the simplified network analysis of a primary science curriculum is examined it will be noticed that there are blank spaces. These have been left quite deliberately. It is quite normal when drawing a network analysis to leave out numbers so that the numbering system on the network is not strictly in sequence. The numbers which have been omitted can be used for any activities which have been left out.

This design study of primary science is a preliminary version and it will need modification before it is ready for use in a classroom. When changes need to be included any series of activities can easily be inserted along one of the blank lines. The blank lines of activities, 4 through to 1005; 7 through to 1020; 9 to 1028 and 13 to 1041 will enable entries to be made of any continuous series of activities. The other blank entries can be used for minor additions.

It must not be assumed that because the network design is set out in a rather formal manner that it must be used only in school periods. Many of the activities could well take place as 'homework', using this phrase to mean out of school. Collections of drawings, pictures or, given suitable conditions, actual objects, can often more appropriately be collected out of school during weekends or holidays. Care of course should be taken to ensure that plants, insects, and other items are collected in a responsible way and according to any regulations designed to protect or preserve the environment.

With this proviso there are many activities which could take place outside school hours. Given support from parents some of the activities which need one-to-one supervision, as for example activities which require heat, can be done at home. Collections of records about rainfall, temperature, etc. could also be extended into holiday periods. There is nothing more discouraging for a pupil than to have a record of rainfall which shows none at all during the school week, and then to see a deluge while at home at the week-end.

In any school area there will be places of interest to visit and full use should be made of them. Full use should be made of museums, collections, libraries and other places in which exhibits are on display. Schools in cities will be better able to enter into activities of this type, whereas schools in country areas, or near the sea will be able to spend more time on activities for which they can more easily find material. Country schools will wish to spend more time on the series of activities involving plant life as they will have more examples ready to hand.

There are many possibilities for links with other subjects, and other teachers could be encouraged to help by linking what they do with the work in science. This can only be done if all have a clear picture of the syllabus, and the network layout should help in the dissemination of this information. In the series of activities about ears and hearing, the music teacher can link in a lot of his work with science. Some pupils who do not enjoy singing, and cannot find much sense in the use of the instruments used in primary schools might find the subject of more interest if their attention can be directed to the scientific aspects of sound production. In this area making simple sound producing instruments like bottle xylophones, one string fiddles, and drums could stir their interest.

The art teachers who can show how colours are mixed in a variety of ways, and carry out experiments in dyeing will find plenty of links with the work in science. In fact many of the activities in the network may well not take place in periods set aside for science, but take place in other periods.

Mention has already been made of the links with music and art, and there are many others which will be revealed by a study of the network. Handwork periods can also be linked into the science scheme of work, and many of the experiments need small items of simple apparatus which can be made in handwork periods.

There are many links with other subjects at primary level and this is one reason why teachers try to implement an integrated working day. The divisions of teaching and learning into subject areas is an artificial one and there is a danger that some important aspect of work may be missed because of it. It may be hoped that when a curriculum is set out in the form of a network diagram the integration of a pupil's work will be made simpler.

In many parts of the network, comparisons will be drawn between the nose, eyes, teeth, hands, etc. of the pupil and those of animals, birds and fish.

This will be made easier if there are pets at school. It is very common to see some animals, birds or fish kept in primary schools. These pets will be invaluable in such comparisons. If it is not possible to keep any pets in school then those at home can be used. Visits to a zoo or farms will also play their part in the build up of experiences.

There are always problems associated with keeping pets in school, and great care should be exercised to ensure that the pets are given the correct environment. In addition to animals and birds, insects such as stick insects, ants or worms can be kept and usually children are fascinated to watch them. I have for some years had bees at school and have been pleasantly surprised by the interest shown by pupils. It is possible for bees to live in an observation hive so that they fly clear of pupils. Bees do not sting unless provoked and there are very few who try to provoke them. As they gather their own food they are very little trouble when holidays arrive, when provision would have to be made for other pets.

At the end of each of the lines of activities there are quite a number of blank spaces. It is here where the curriculum of the primary school links with that of the secondary school that much thought will need to be given to ensure that the two knit together so that pupils can carry on with their course with the minimum of disruption. The blank spaces will enable this area of possible overlap to be studied and the appropriate activities inserted.

There are some aspects of the teaching of science which are peculiar to the subject. There is a need for emphasis on conservation of all living things. Pupils should be encouraged to take great care not to add to their collections any plant or animal which is protected. Whenever possible they should be encouraged to observe and record, rather than to interfere. In the same vein it is not only necessary to preserve wildlife, but also to ensure the safety of pupils. Warnings should be given about dangerous plants, animals and chemicals. These good habits need to be inculcated at the earliest stages of learning.

It can quickly be seen that many of the activities in the science curriculum are of an extended nature, and so may continue over a period. The aim should be to match the length of the activity to the developmental stage of the pupil. Younger pupils having a short attention span will need shorter activities, as they develop they can engage in the more extended exercises.

As in some of the problem-solving activities there are a number of open-ended activities to which there may be no single correct answer. This type of exploratory, open-ended activity is very valuable. Pupils should be encouraged to seek explanations when they have completed a series of observations, and not just look for the 'correct' answer.

Diagram of layout of network of science curriculum

Fig. 39, pt 1, page 49	Fig. 39, pt 2, page 50
Fig. 39, pt 3, page 51	Fig. 39, pt 4, page 52
Fig. 39, pt 5, page 53	Fig. 39, pt 6, page 54
Fig. 39, pt 7, page 55	Fig. 39, pt 8, page 56

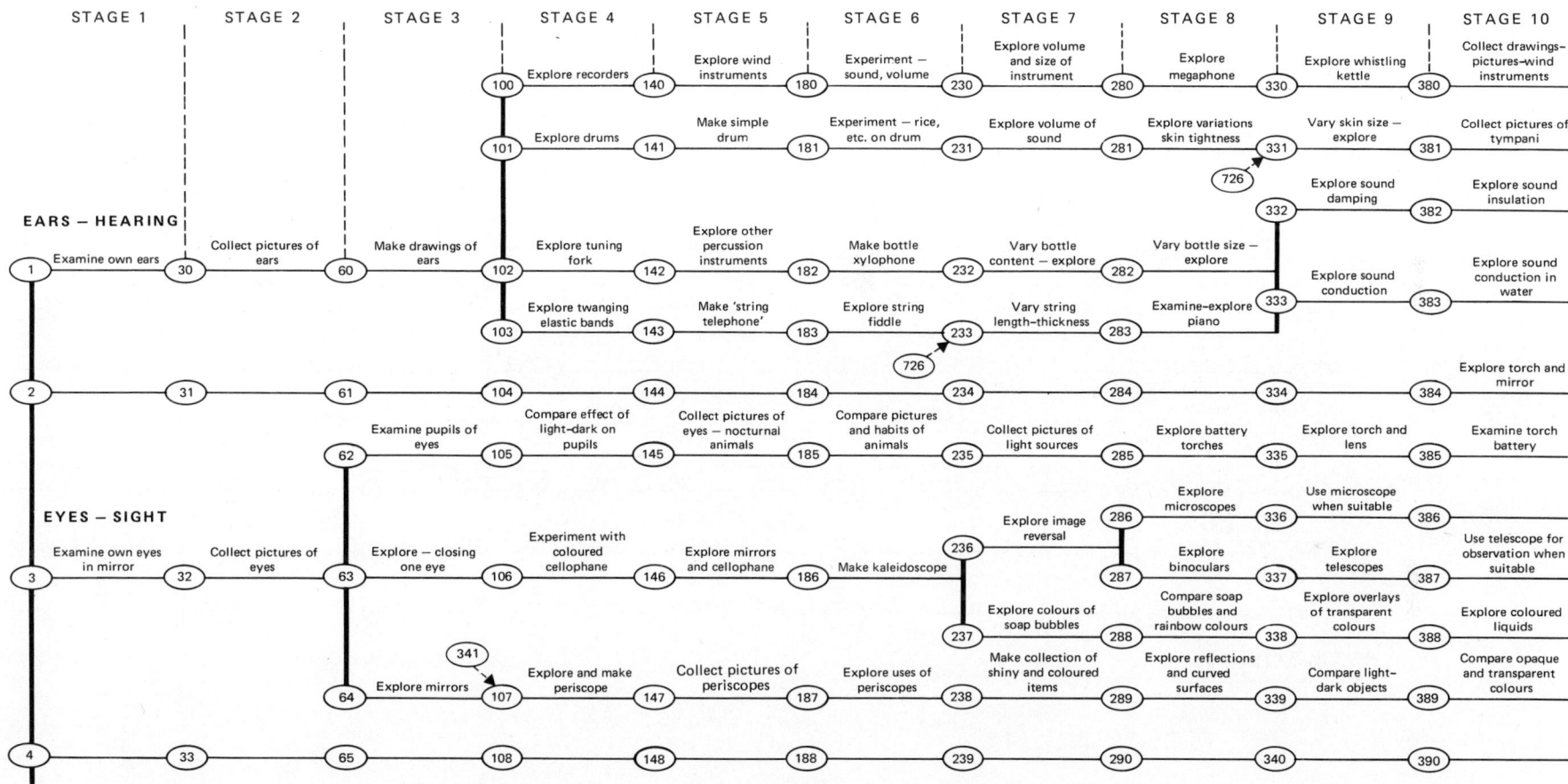

Fig. 39. *A modified network analysis of science curriculum for primary/middle school (9–13 years). The numbered nodes and dotted lines (e.g. (726) - ➔) indicate a dependence upon an activity on the mathematics curriculum.*

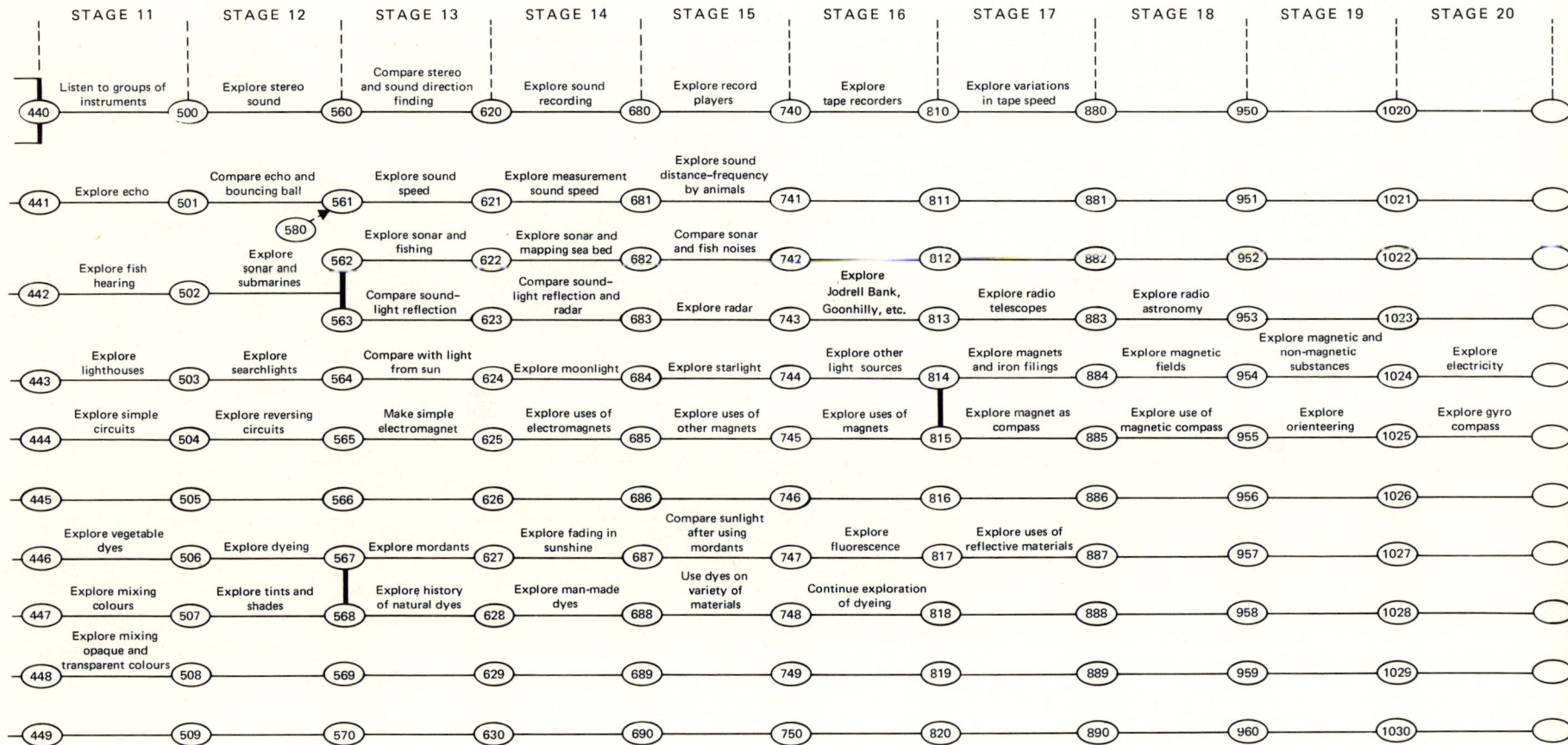

STAGE 11	STAGE 12	STAGE 13	STAGE 14	STAGE 15	STAGE 16	STAGE 17	STAGE 18	STAGE 19	STAGE 20

Listen to groups of instruments (440) — Explore stereo sound (500) — Compare stereo and sound direction finding (560) — Explore sound recording (620) — Explore record players (680) — Explore tape recorders (740) — Explore variations in tape speed (810) — (880) — (950) — (1020) —

Explore echo (441) — Compare echo and bouncing ball (501) — Explore sound speed (561) — Explore measurement sound speed (621) — Explore sound distance–frequency by animals (681) — (741) — (811) — (881) — (951) — (1021)

(580)

Explore fish hearing (442) — Explore sonar and submarines (502) — Explore sonar and fishing (562) — Explore sonar and mapping sea bed (622) — Compare sonar and fish noises (682) — (742) — (812) — (882) — (952) — (1022)

(563) Compare sound–light reflection — Compare sound–light reflection and radar (623) — Explore radar (683) — Explore Jodrell Bank, Goonhilly, etc. (743) — Explore radio telescopes (813) — Explore radio astronomy (883) — (953) — (1023)

Explore lighthouses (443) — Explore searchlights (503) — Compare with light from sun (564) — Explore moonlight (624) — Explore starlight (684) — Explore other light sources (744) — Explore magnets and iron filings (814) — Explore magnetic fields (884) — Explore magnetic and non-magnetic substances (954) — Explore electricity (1024)

Explore simple circuits (444) — Explore reversing circuits (504) — Make simple electromagnet (565) — Explore uses of electromagnets (625) — Explore uses of other magnets (685) — Explore uses of magnets (745) — Explore magnet as compass (815) — Explore use of magnetic compass (885) — Explore orienteering (955) — Explore gyro compass (1025)

(445) — (505) — (566) — (626) — (686) — (746) — (816) — (886) — (956) — (1026)

Explore vegetable dyes (446) — Explore dyeing (506) — Explore mordants (567) — Explore fading in sunshine (627) — Compare sunlight after using mordants (687) — Explore fluorescence (747) — Explore uses of reflective materials (817) — (887) — (957) — (1027)

Explore mixing colours (447) — Explore tints and shades (507) — Explore history of natural dyes (568) — Explore man-made dyes (628) — Use dyes on variety of materials (688) — Continue exploration of dyeing (748) — (818) — (888) — (958) — (1028)

Explore mixing opaque and transparent colours (448) — (508) — (569) — (629) — (689) — (749) — (819) — (889) — (959) — (1029)

(449) — (509) — (570) — (630) — (690) — (750) — (820) — (890) — (960) — (1030)

Fig. 39 (part 2).

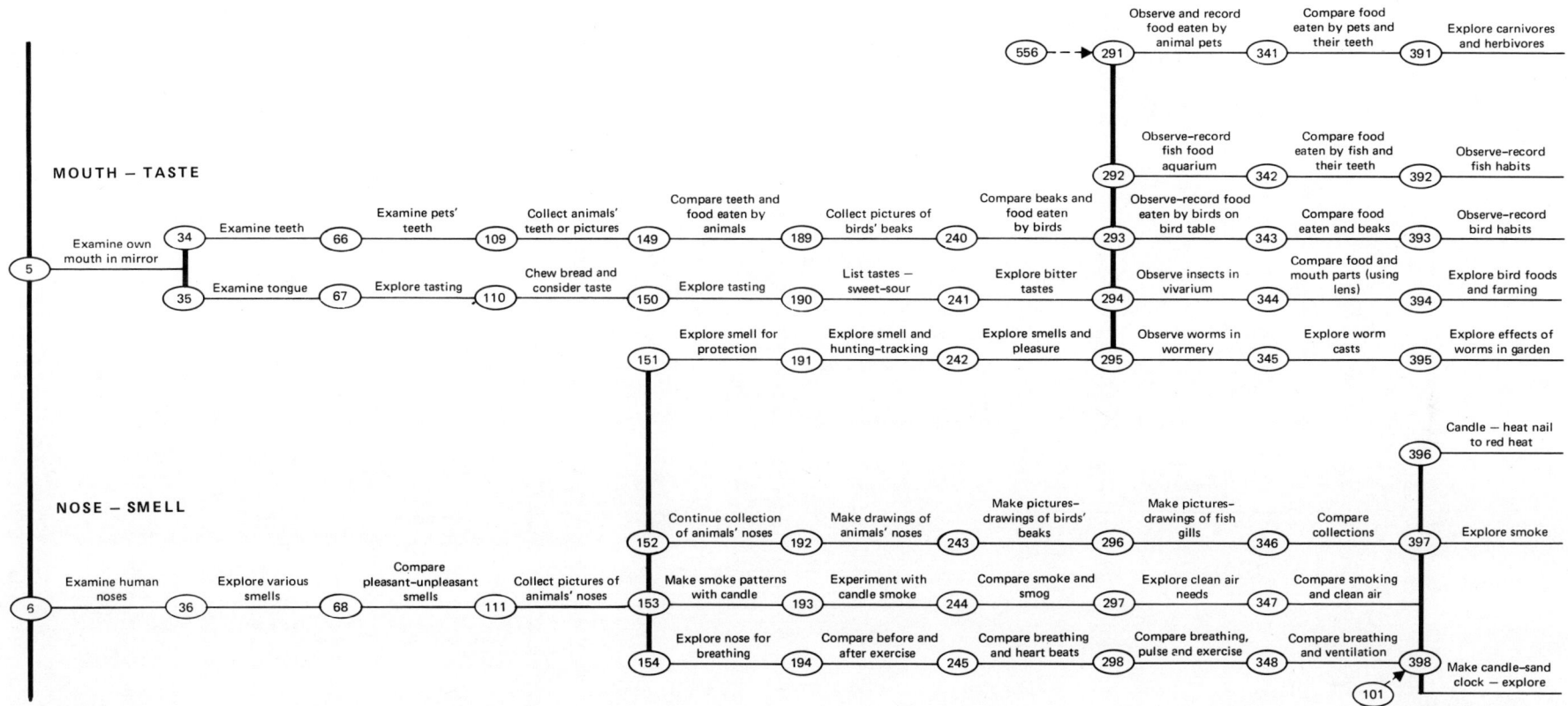

Fig. 39 (part 3).

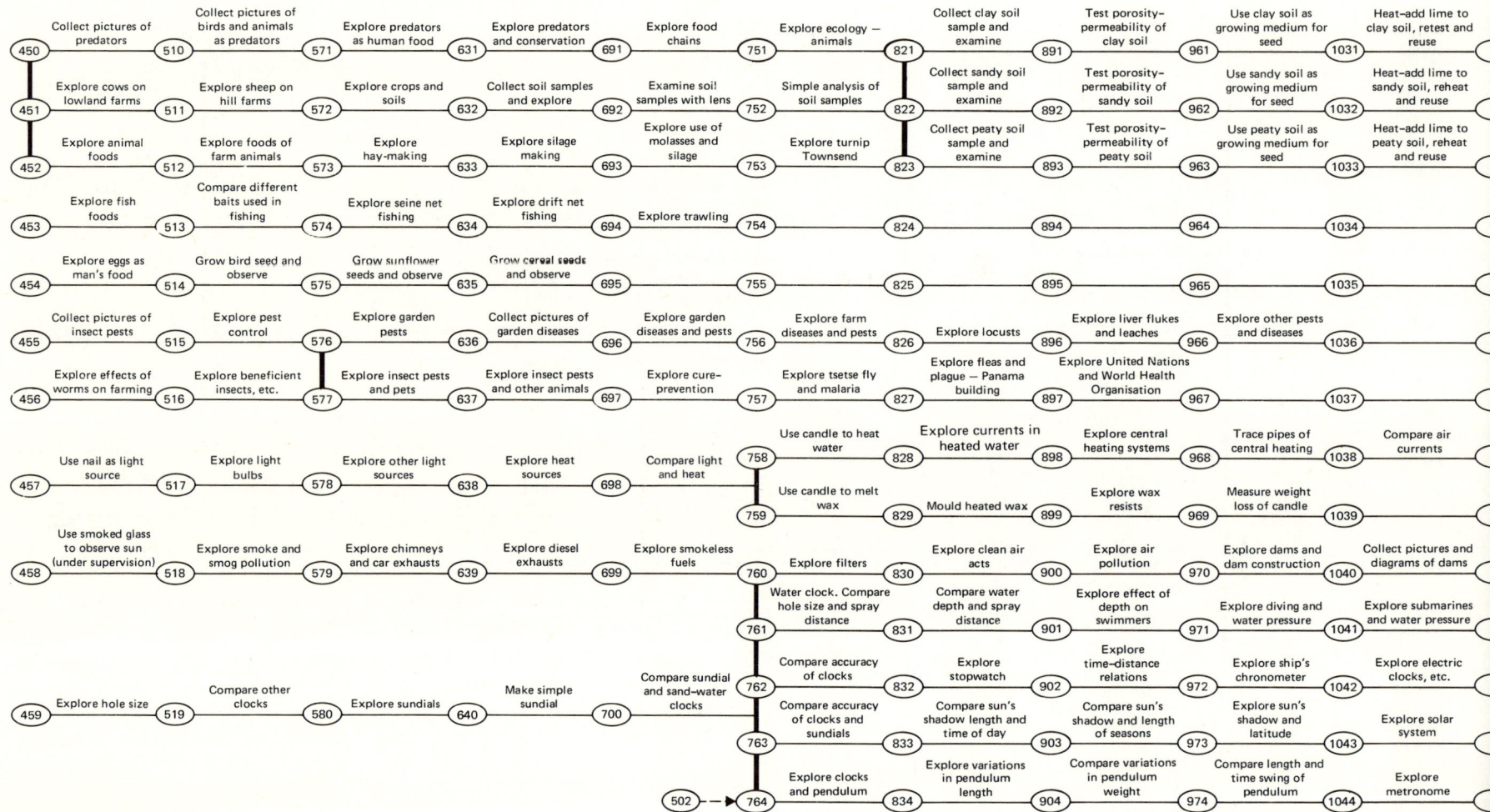

450 Collect pictures of predators	510 Collect pictures of birds and animals as predators	571 Explore predators as human food	631 Explore predators and conservation	691 Explore food chains	751 Explore ecology — animals	821 Collect clay soil sample and examine	891 Test porosity-permeability of clay soil	961 Use clay soil as growing medium for seed	1031 Heat–add lime to clay soil, retest and reuse
451 Explore cows on lowland farms	511 Explore sheep on hill farms	572 Explore crops and soils	632 Collect soil samples and explore	692 Examine soil samples with lens	752 Simple analysis of soil samples	822 Collect sandy soil sample and examine	892 Test porosity-permeability of sandy soil	962 Use sandy soil as growing medium for seed	1032 Heat–add lime to sandy soil, reheat and reuse
452 Explore animal foods	512 Explore foods of farm animals	573 Explore hay-making	633 Explore silage making	693 Explore use of molasses and silage	753 Explore turnip Townsend	823 Collect peaty soil sample and examine	893 Test porosity-permeability of peaty soil	963 Use peaty soil as growing medium for seed	1033 Heat–add lime to peaty soil, reheat and reuse
453 Explore fish foods	513 Compare different baits used in fishing	574 Explore seine net fishing	634 Explore drift net fishing	694 Explore trawling	754	824	894	964	1034
454 Explore eggs as man's food	514 Grow bird seed and observe	575 Grow sunflower seeds and observe	635 Grow cereal seeds and observe	695	755	825	895	965	1035
455 Collect pictures of insect pests	515 Explore pest control	576 Explore garden pests	636 Collect pictures of garden diseases	696 Explore garden diseases and pests	756 Explore farm diseases and pests	826 Explore locusts	896 Explore liver flukes and leaches	966 Explore other pests and diseases	1036
456 Explore effects of worms on farming	516 Explore beneficient insects, etc.	577 Explore insect pests and pets	637 Explore insect pests and other animals	697 Explore cure-prevention	757 Explore tsetse fly and malaria	827 Explore fleas and plague — Panama building	897 Explore United Nations and World Health Organisation	967	1037
457 Use nail as light source	517 Explore light bulbs	578 Explore other light sources	638 Explore heat sources	698 Compare light and heat	758 Use candle to heat water	828 Explore currents in heated water	898 Explore central heating systems	968 Trace pipes of central heating	1038 Compare air currents
					759 Use candle to melt wax	829 Mould heated wax	899 Explore wax resists	969 Measure weight loss of candle	1039
458 Use smoked glass to observe sun (under supervision)	518 Explore smoke and smog pollution	579 Explore chimneys and car exhausts	639 Explore diesel exhausts	699 Explore smokeless fuels	760 Explore filters	830 Explore clean air acts	900 Explore air pollution	970 Explore dams and dam construction	1040 Collect pictures and diagrams of dams
					761 Water clock. Compare hole size and spray distance	831 Compare water depth and spray distance	901 Explore effect of depth on swimmers	971 Explore diving and water pressure	1041 Explore submarines and water pressure
459 Explore hole size	519 Compare other clocks	580 Explore sundials	640 Make simple sundial	700 Compare sundial and sand-water clocks	762 Compare accuracy of clocks	832 Explore stopwatch	902 Explore time-distance relations	972 Explore ship's chronometer	1042 Explore electric clocks, etc.
					763 Compare accuracy of clocks and sundials	833 Compare sun's shadow length and time of day	903 Compare sun's shadow and length of seasons	973 Explore sun's shadow and latitude	1043 Explore solar system
502					764 Explore clocks and pendulum	834 Explore variations in pendulum length	904 Compare variations in pendulum weight	974 Compare length and time swing of pendulum	1044 Explore metronome

Fig. 39 (part 4).

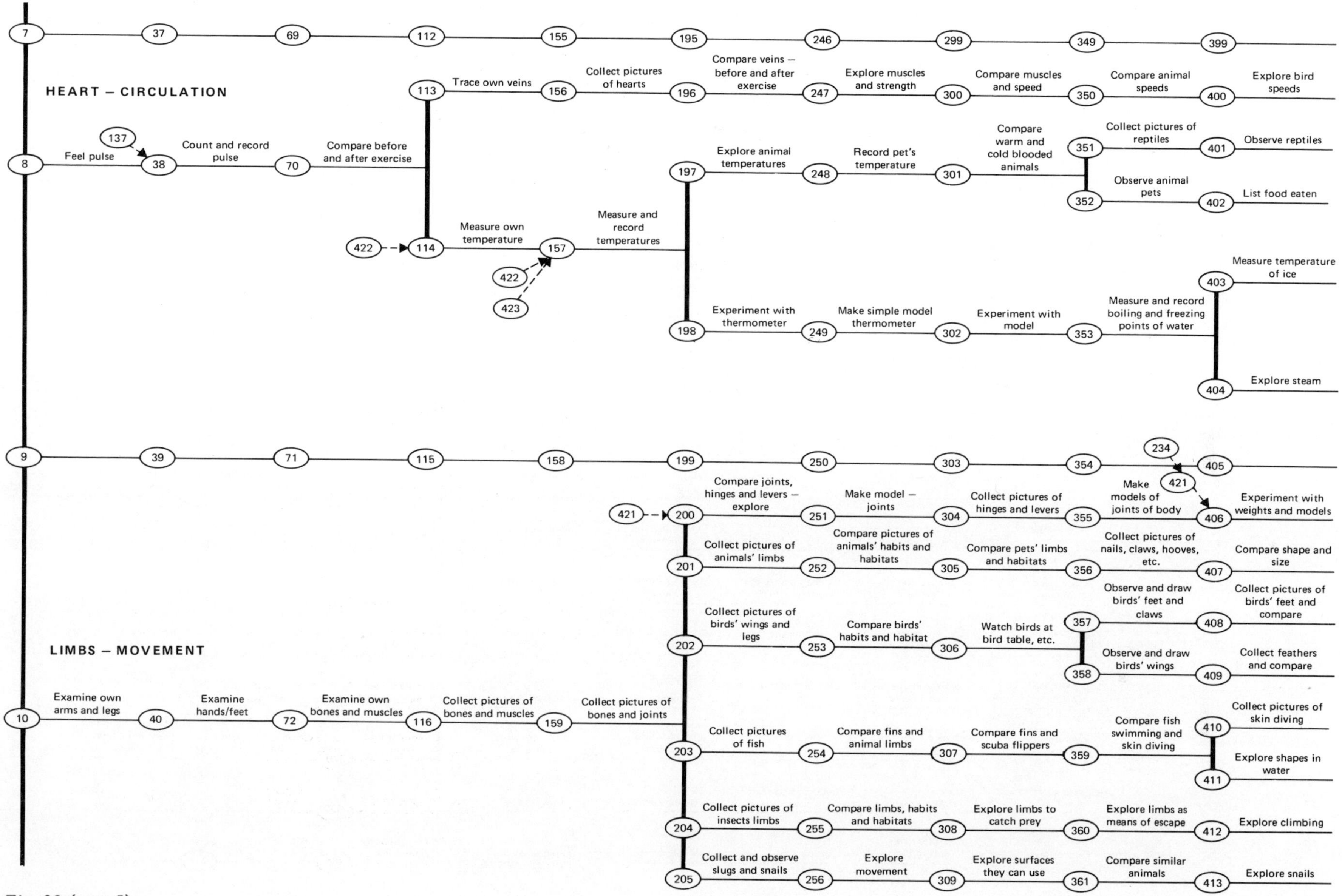

HEART – CIRCULATION

Trace own veins

Collect pictures of hearts

Compare veins – before and after exercise

Explore muscles and strength

Compare muscles and speed

Compare animal speeds

Explore bird speeds

Feel pulse

Count and record pulse

Compare before and after exercise

Explore animal temperatures

Record pet's temperature

Compare warm and cold blooded animals

Collect pictures of reptiles

Observe reptiles

Observe animal pets

List food eaten

Measure own temperature

Measure and record temperatures

Experiment with thermometer

Make simple model thermometer

Experiment with model

Measure and record boiling and freezing points of water

Measure temperature of ice

Explore steam

LIMBS – MOVEMENT

Compare joints, hinges and levers – explore

Make model – joints

Collect pictures of hinges and levers

Make models of joints of body

Experiment with weights and models

Collect pictures of animals' limbs

Compare pictures of animals' habits and habitats

Compare pets' limbs and habitats

Collect pictures of nails, claws, hooves, etc.

Compare shape and size

Collect pictures of birds' wings and legs

Compare birds' habits and habitat

Watch birds at bird table, etc.

Observe and draw birds' feet and claws

Collect pictures of birds' feet and compare

Observe and draw birds' wings

Collect feathers and compare

Examine own arms and legs

Examine hands/feet

Examine own bones and muscles

Collect pictures of bones and muscles

Collect pictures of bones and joints

Collect pictures of fish

Compare fins and animal limbs

Compare fins and scuba flippers

Compare fish swimming and skin diving

Collect pictures of skin diving

Explore shapes in water

Collect pictures of insects limbs

Compare limbs, habits and habitats

Explore limbs to catch prey

Explore limbs as means of escape

Explore climbing

Collect and observe slugs and snails

Explore movement

Explore surfaces they can use

Compare similar animals

Explore snails

Fig. 39 (part 5).

53

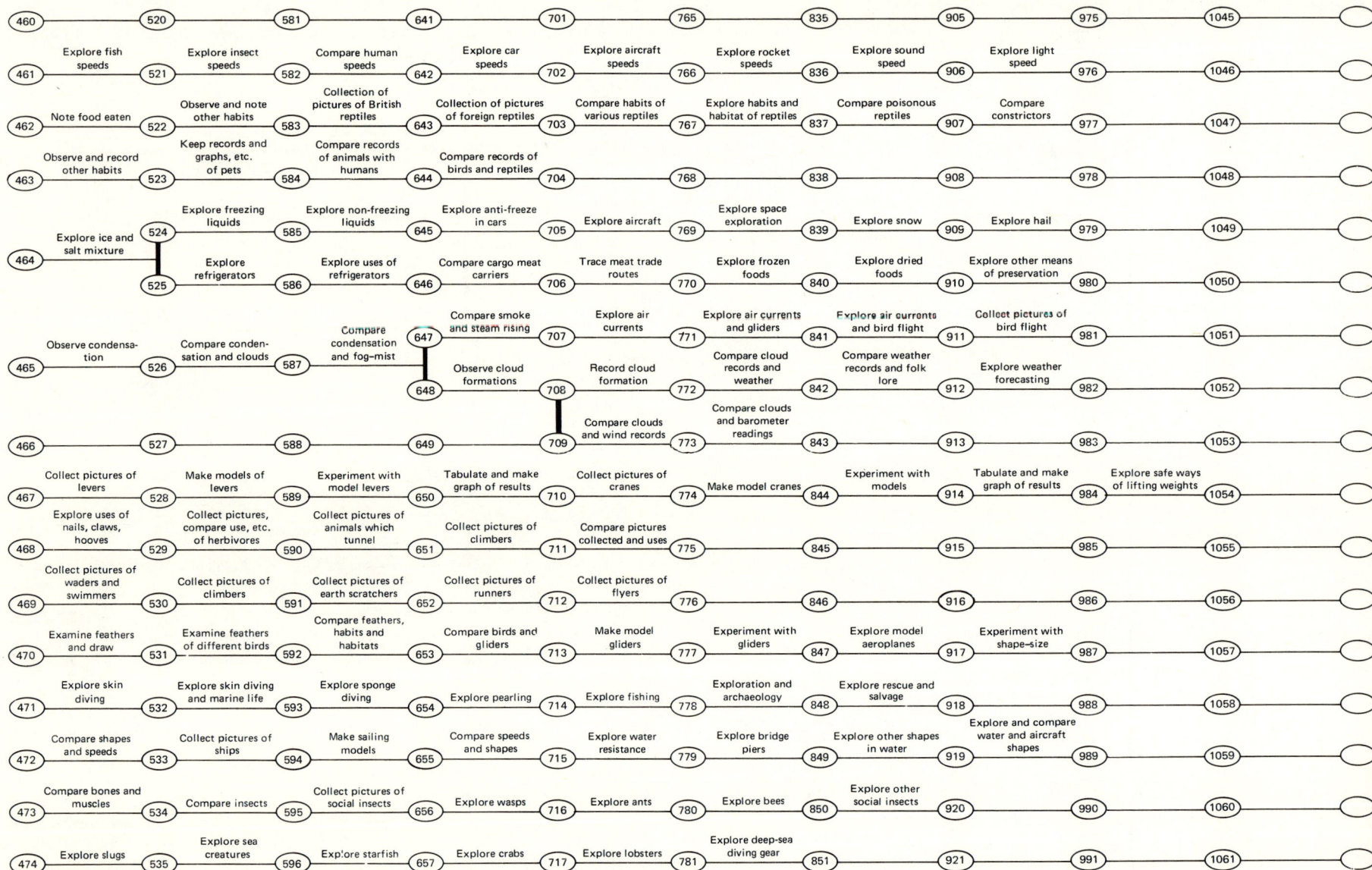

460 — 520 — 581 — 641 — 701 — 765 — 835 — 905 — 975 — 1045

461 Explore fish speeds — 521 Explore insect speeds — 582 Compare human speeds — 642 Explore car speeds — 702 Explore aircraft speeds — 766 Explore rocket speeds — 836 Explore sound speed — 906 Explore light speed — 976 — 1046

462 Note food eaten — 522 Observe and note other habits — 583 Collection of pictures of British reptiles — 643 Collection of pictures of foreign reptiles — 703 Compare habits of various reptiles — 767 Explore habits and habitat of reptiles — 837 Compare poisonous reptiles — 907 Compare constrictors — 977 — 1047

463 Observe and record other habits — 523 Keep records and graphs, etc. of pets — 584 Compare records of animals with humans — 644 Compare records of birds and reptiles — 704 — 768 — 838 — 908 — 978 — 1048

464 Explore ice and salt mixture
— 524 Explore freezing liquids — 585 Explore non-freezing liquids — 645 Explore anti-freeze in cars — 705 Explore aircraft — 769 Explore space exploration — 839 Explore snow — 909 Explore hail — 979 — 1049
— 525 Explore refrigerators — 586 Explore uses of refrigerators — 646 Compare cargo meat carriers — 706 Trace meat trade routes — 770 Explore frozen foods — 840 Explore dried foods — 910 Explore other means of preservation — 980 — 1050

465 Observe condensation — 526 Compare condensation and clouds — 587 Compare condensation and fog-mist
— 647 Compare smoke and steam rising — 707 Explore air currents — 771 Explore air currents and gliders — 841 Explore air currents and bird flight — 911 Collect pictures of bird flight — 981 — 1051
— 648 Observe cloud formations — 708 Record cloud formation — 772 Compare cloud records and weather — 842 Compare weather records and folk lore — 912 Explore weather forecasting — 982 — 1052

466 — 527 — 588 — 649 — 709 Compare clouds and wind records — 773 Compare clouds and barometer readings — 843 — 913 — 983 — 1053

467 Collect pictures of levers — 528 Make models of levers — 589 Experiment with model levers — 650 Tabulate and make graph of results — 710 Collect pictures of cranes — 774 Make model cranes — 844 Experiment with models — 914 Tabulate and make graph of results — 984 Explore safe ways of lifting weights — 1054

468 Explore uses of nails, claws, hooves — 529 Collect pictures, compare, use, etc. of herbivores — 590 Collect pictures of animals which tunnel — 651 Collect pictures of climbers — 711 Compare pictures collected and uses — 775 — 845 — 915 — 985 — 1055

469 Collect pictures of waders and swimmers — 530 Collect pictures of climbers — 591 Collect pictures of earth scratchers — 652 Collect pictures of runners — 712 Collect pictures of flyers — 776 — 846 — 916 — 986 — 1056

470 Examine feathers and draw — 531 Examine feathers of different birds — 592 Compare feathers, habits and habitats — 653 Compare birds and gliders — 713 Make model gliders — 777 Experiment with gliders — 847 Explore model aeroplanes — 917 Experiment with shape–size — 987 — 1057

471 Explore skin diving — 532 Explore skin diving and marine life — 593 Explore sponge diving — 654 Explore pearling — 714 Explore fishing — 778 Exploration and archaeology — 848 Explore rescue and salvage — 918 — 988 — 1058

472 Compare shapes and speeds — 533 Collect pictures of ships — 594 Make sailing models — 655 Compare speeds and shapes — 715 Explore water resistance — 779 Explore bridge piers — 849 Explore other shapes in water — 919 Explore and compare water and aircraft shapes — 989 — 1059

473 Compare bones and muscles — 534 Compare insects — 595 Collect pictures of social insects — 656 Explore wasps — 716 Explore ants — 780 Explore bees — 850 Explore other social insects — 920 — 990 — 1060

474 Explore slugs — 535 Explore sea creatures — 596 Explore starfish — 657 Explore crabs — 717 Explore lobsters — 781 Explore deep-sea diving gear — 851 — 921 — 991 — 1061

Fig. 39 (part 6).

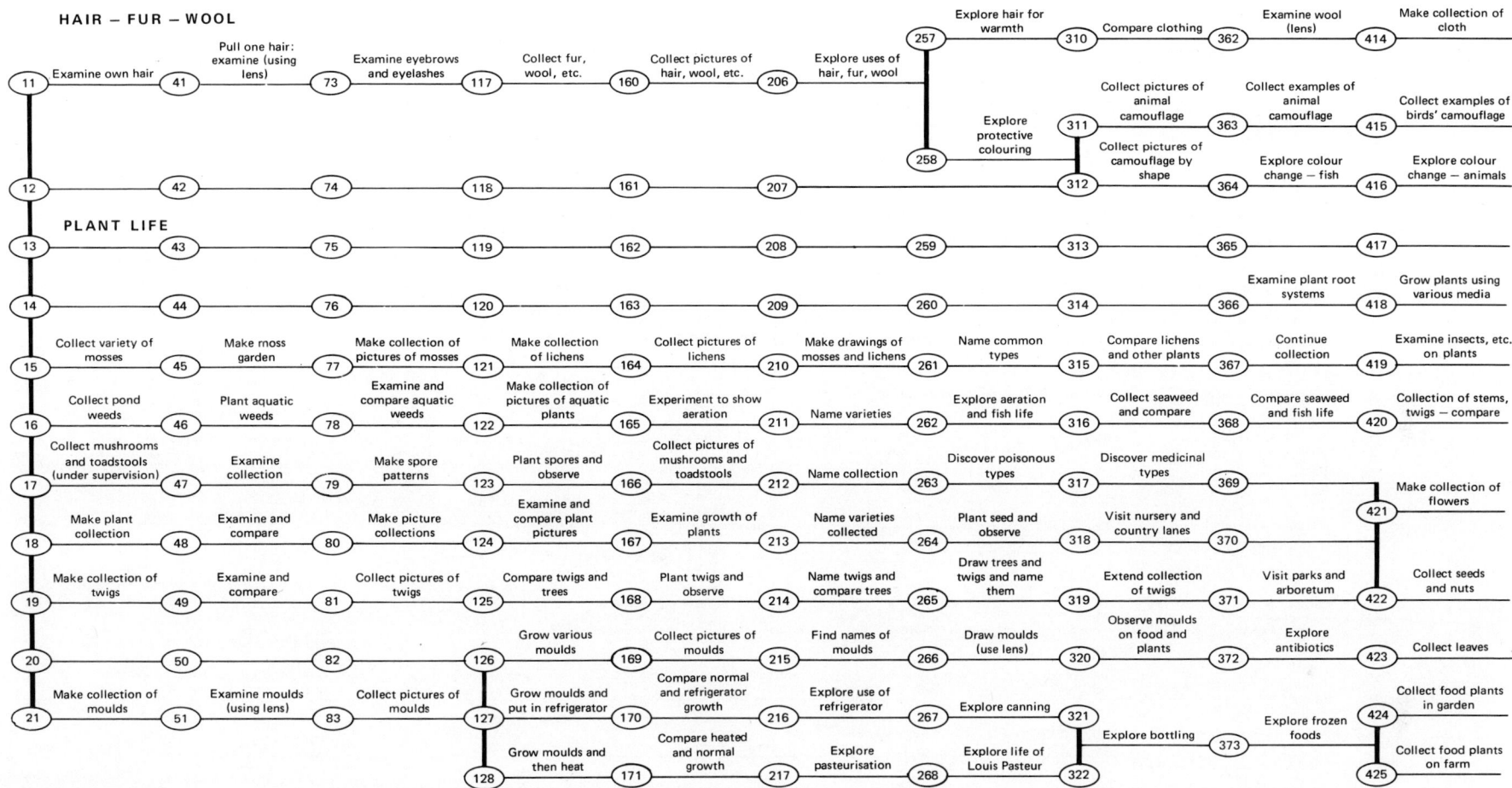

Fig. 39 (part 7).

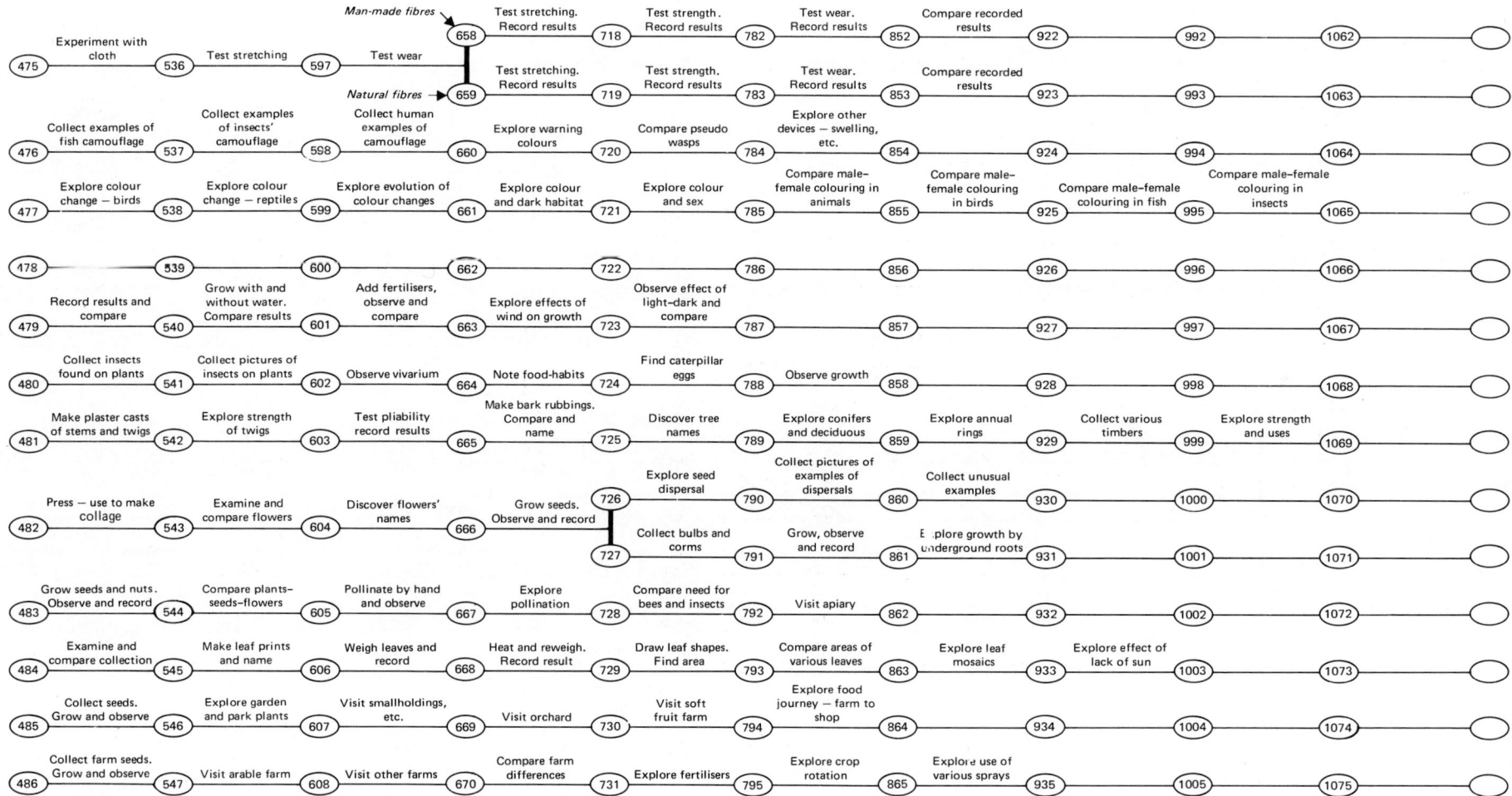

Fig. 39 (part 8).

Art and craft

In recent years primary school teachers have been faced with an ever increasing amount of information about art and craft. Each year more books are published explaining how a bewildering variety of materials can be shaped into finished works of art. At the same time similar information is becoming available for other subjects so that the class teacher is hard pressed to keep informed. Subject barriers are losing definition fast, and art and craft are used more and more as tools in fields such as mathematics, geography, history and language.

Meanwhile traditional teaching methods are changing so that the child is receiving more attention as an individual. The art lesson is now no longer the only way in which a pupil is allowed to pursue his creative course once a week. With the integrated day art and craft work may be undertaken in the class for a large part of the week by children singly, or in small groups. This has meant that it has been increasingly difficult for the teacher to follow the traditional art syllabus, which has thus been discarded, and very often the materials have simply been provided so that children can 'explore' their properties. Very often exploration has continued as a child passes from teacher to teacher with the pupil only gaining haphazard information about specific techniques.

Sometimes the teacher has evolved a record-keeping system for tracing a child's passage through subjects such as mathematics and English, and this record is used by subsequent teachers to provide an organised body of knowledge, one concept building upon another. This record helps the teacher know precisely how much work has been covered in a certain field, and also when to provide the student with a new skill or encouragement to stimulate progress.

When we come to art and craft however it seems that experience gained in other fields is not applied, as it is felt that creativity cannot be organised. It is certainly true that no one can be taught to be creative, but it is also true that the techniques of the artist can be taught. Many children may not discover for themselves how to make a coil pot from a lump of clay, or how to make a monoprint from a finger painting but once they have been helped with the basic activity they can produce work which is truly original.

It is not usually the aim of every school to turn out artists, so it is as well to discover why art and craft have such important rôles in the primary school. Probably there are two main rôles. The first is to improve a child's perception and dexterity through the handling of media. The second provides an outlet for creativity, which may be a kind of therapy for the less able child. Obviously the two branches of art and craft are interdependant, but the latter does depend much more on the former.

The network presented is intended to provide the teacher with a skeleton of information, simple enough for him to absorb readily, but providing enough references for detail to be obtained if required. When it is considered that the information concerning art and craft is only a small part of the total knowledge required by a non-specialist primary teacher to educate a group of children, it becomes evident that some such system is required. The aim is that this logical arrangement of creative materials and practical techniques should fill the gap left by the gradual disappearance of the traditional syllabus until such a time that a better system can be developed to replace it. The network is thus a support for those teachers who may have been dismayed by the huge quantity of information about art and craft and found themselves unable to make efficient use of the literature and resources available. It is intended as a guide for the non-specialist art teacher, so that he may provide an organised, but truly imaginative basis for art and craft in the primary school. The extra numbers on the network lines refer the teacher to selected reference books (see p. 61).

It will have been noticed that the network has not been linked in any way either with the mental or chronological ages of the pupils engaged in the operations. Also there is no mention of infant or junior stages in the scheme. The network has been arranged so that the operations on the left of the diagram are easier than those on the right. The rate at which a child progresses from one skill to another depends entirely on his own ability and not on the fact that he has now reached a certain age and so can undertake a certain technique. This automatically places the onus on the teacher to decide when a pupil has enough skill or perception to progress to a more advanced field of knowledge. At the same time although the network suggests that certain techniques follow on from simpler ones, this is sometimes an arbitrary matter and it is intended that teachers should alter and annotate the network to suit their own personal way of working. The arrangement of the information is such that new material can be included easily and unwanted techniques discarded.

Once a specific technique is required for use in a class area it is not the intention that the network should tell the teacher the exact way it should be taught. The intention is that the accompanying list of reference books should be consulted for ideas, but other sources or approaches may be known, in which case that information could be used instead. Actual page numbers have been deliberately left out of the references as many of the books devote large areas to similar techniques. Choice should be a matter of the teacher's taste and also depends on the reference books at his disposal.

It is not suggested, by the way, that every pupil should commence with the top stratum, i.e. balsa, and work his way through the entire section before beginning the layer below, i.e. plastic. The network is not to be read like the pages of a book, and starts can be made in each section when the teacher decides that a new experience in a different material is required. This means that any one pupil may have started work with all the materials after his first year in school, although he may have progressed further in the techniques of one section than another, because of his own aptitude or the availability of those particular materials. It may be that some teachers prefer to devote whole blocks of time to a particular material throughout the school year and ignore the others, or it may be that all the materials can be available all the year round for use as required. Either approach is valid and depends entirely on the staff or school concerned.

The network is simply an attempt to connect certain related techniques in a logically based sequence so that the pupil can at least have had some relevant experience to draw upon when faced with his next stimulus. Sometimes it is a matter of choice as to which operation is most valuable next. Here the network divides to provide parallels, or if the teacher feels it necessary the pupil can be fed back to the starting point to provide enrichment for a particular idea.

With such an arrangement it is possible for the teacher to organise the pupil so that he gains the maximum experience where he is weakest, but it is also possible to digress to loosely related ideas without losing the overall aim involved. There is no necessity for the student always to work across the chart from left to right, as many of the activities can be explored a second time at a deeper level without enjoyment being lost.

The choice of headings for each section of the network has been made for the sake of simplicity and clarity, and not because the validity or existence of particular materials has not been recognised. Each section is intended to cover a group of materials with broadly similar properties. Thus although one section is entitled 'clay' it is intended to include materials such as modelling clay, dough and plaster of Paris which behave in a similar manner, or are used in conjunction with clay.

A word of explanation about each section would probably be useful here.

Balsa. This title was chosen for a section concerning wood as it was felt that balsa wood required less physical strength than any other wood to manage and so was suitable for a child to handle. Other harder woods, such as pine, lend themselves to many of the techniques listed and if it was found that individual pupils could cope with the extra skill and strength required, these would serve in the place of balsa.

The skills needed for the use of wood vary from those of the most basic kind where a pupil arranges or assembles prepared shapes or found objects, for example tree roots or driftwood, without actually having to cut or shape them in any way, to more complex tasks where a tool or set of tools is used. The use of most tools has to be taught if damage to the tool and user is to be avoided. Once skill with a single tool is gained from activities that emphasise its rôle, this skill can be used singly in creative activities such as model making, or in combination with other techniques, equipment and media, for example making wood printing blocks or picture frames.

Plastic. The smallest section of the network, this is probably the one which needs revision most. The term 'plastic' is intended to include all kinds of commercial packaging, as well as the expanded polystyrene tiles, etc., much used in schools.

The simplest use of plastic is to assemble cartons or prepared pieces of polystyrene ceiling tile into constructions or arrangements using a suitable adhesive. No attempt is made by the pupil to cut the plastic in any way. The latter kind of construction can be developed into the more complex field of construction kits. A more demanding technique developed from the basic activity is cutting the plastic to shape before assembling the components. Here scissors or a knife can be used depending on the child and the plastic used. Granular expanded polystyrene and the more compact plastic packaging behave very differently when they are cut, hence the choice of tool is important.

The former can be most readily cut with a hot wire cutting tool, but this requires some dexterity, and also certain kinds give off poisonous fumes and so should be avoided. Carving is a more difficult technique than straightforward cutting and may involve the use of still more sophisticated tools such as files and rasps. No mention has been made of moulding plastic, as this normally requires the use of heat, and the chemical nature of some packaging may be such that dangerous fumes are given off. This activity should therefore be avoided with young children.

Clay. This section has been expanded to include modelling clay and dough, which have similar properties to clay when unfired. Plaster of Paris is also included under this heading, although it might well have a heading of its own, or be included with metal where it is much used with armatures.

The most basic experience with clay is the preparatory handling of the material to discover some of its properties. At the earliest stage it is useful to decide whether what is made is going to be preserved by firing or dried in the atmosphere. The child should learn ways of removing air from the lump of clay as early as possible if it is to be fired. The skill of wedging once learned can then be applied to any stage. Even though wedging may be ignored, shaping the clay into simple solid shapes such as balls, cylinders, and cubes provides the pupil with some useful skills which do not require the use of tools. Nearly all the shaping activities that follow develop from altering these basic solid shapes. Thus a thumb pot is a ball pierced by the thumbs, a slab is a flattened cube, a cone is a distorted cylinder and a coil pot is a very long cylinder wound round in a spiral. Many of these activities do not require tools, but as complexity increases cheese wire, roller and modelling tools should be introduced. The hardest activities are those where fine distinctions and accurate measurements are needed, as well as the ability to handle several tools. The construction of a slab pot comes in the latter category hence its place on the right of the network.

If a kiln is available in the school, some techniques concerned with firing may be learned in conjunction with the activities mentioned previously. The decorative skills, requiring the use of oxides, slips and glazes, also vary in complexity and so have been arranged on the network so that they can be absorbed in parallel with clay techniques. The use of plaster of Paris is included in the middle of the network as mixing is fairly difficult, and anyway plaster of Paris moulds are needed to shape some of the harder clay vessels, or as a matrix for tile or glass mosaics.

Metal. Aluminium foil provides good early experience in the properties of metal, without requiring physical strength in the pupil, or the use of dangerous and sophisticated tools. To the foil can be added soft iron wire and metal food containers in the early stages of the network; tin plate and other soft metals can be used later on if adequate supervision is given and the dangers recognised.

The metal section of the network divides initially into two parallel series of activities. Children can explore construction work with containers, electrical components, insulated wire, milk tops, etc. using impact adhesives and working without tools. At the same time different experience can be gained from work with pipe cleaners. Both layers then develop into techniques using simple tools like the hammer and punch, or pliers. The upper layer is more concerned with the use of wire and perforated metal, which can be bent and twisted easily. Plaster of Paris mixing is mentioned here as the wire makes excellent armatures for sculptures and models. The lower layer concentrates on metal plates and the ways in which they can be affected by cutting or bending. Perhaps the most difficult metalwork activity in the primary school is enamelling. It is therefore placed on the right of the chart as it involves the use of many of the skills, i.e. cutting and smoothing, previously experienced.

59

Card and paper. Card and paper have been given a section of their own for activities involving the use of these materials in their own right rather than as a background for other media. The use of coloured paper or card is not mentioned separately, although one of the references gives some suggestions.

As with other media the simplest activity using paper and card is collage work with shapes cut or torn previously by an adult, or construction with commercial packaging, such as boxes and tubes. A new concept is introduced if the medium is altered in some way, by cutting, tearing, or bending. The horizontal divisions in the network at this stage are decided by the initial shape of the paper or card concerned. Thus we have boxes and tubes, flat polyhedra, and various forms of paper sheet. Tools are not complicated in this section; scissors, knife and straight edge are all that is required. Experience with different adhesives can be given here. On the right of the network have been placed those activities which require some knowledge of how paper and card behave, such as constructions which require the strength of a fold or tube and skills which take advantage of the transparent or translucent nature of some papers. Activities such as mask-making, paper flowers and mobiles are not the only ones that involve use of this medium and many others known to teachers can be inserted here. Many of the techniques are equally applicable to other media so cross references have been provided. Work on weaving paper strips could be carried out alongside similar work with textiles, or masks could be made out of cardboard as well as clay or wood.

Solid colour. Incorporated in the term 'solid colour' are materials such as pencils, chalks, crayons and charcoal. These media have been separated from liquid colour due to the fact that many of their properties are different, even though many of the concepts required for their use are common.

The first use of solid colour is in activities designed to emphasise the way in which a medium can make different marks on a background. In this area come the first scribbles of the small child, the use of point and edge, smudging, and controlled movements. The chart has a triple division after the initial stage. The paths are concerned with mastering the basic information about colour and perception, learning techniques which increase co-ordination, or employing a simple skill which allows use to be made of another object's inherent or applied texture, i.e. rubbings. The three layers of the diagram continue with activities intended to increase perception and co-ordination further by using tools such as stencils, templates and simple drawing instruments. Sophisticated concepts such as proportion and perspective may not be understood by many children in the primary school, but have a place on the right of the network for those gifted children who might need such information.

Textiles. Basketry has been included with textiles as some of the activities are very similar. Leatherwork might also be included in some sections although it is not mentioned by name.

The textile activities vary according to whether the student is to make his own cloth before working with it, or whether ready-made cloth is used. The network has an initial division to allow the pupil to pursue either course, or both.

If material already woven is going to be used, collage with prepared pieces of fabric is a basic starting point. This stage may last a long time as the variety of texture and colour provides much experience without the pupil having to cut or sew. Once some basic exploration of the nature of textiles has been done, the use of scissors, needle and thread can be introduced. Simple gluing of ready-cut fabric can develop into cutting, gluing, and then sewing. The simple stitches can be explored for their usefulness as well as their decorative qualities, and finally a number of skills can be used to construct clothing or produce decorative embroidered designs.

If textiles are to be made, carding and spinning

will be the initial skills for a child to learn. Weaving or plaiting of the thread follows, or the dyeing process if preferred. The techniques of weaving are many and varied, but some are certainly more complicated than others, so these have been arranged in approximate order of the skill required. Dyeing also covers a large field, so the activities including such potential dangers as hot wax, or harmful chemicals such as bleach have been arranged in order of possible difficulty. Plaiting and finger knitting develop into harder skills requiring the use of tools such as the needle and crochet hook.

Once the skills from both divisions of the textile section have been gained there is of course no reason why they should not be used together. Many techniques can be used in parallel long before all the other activities have been covered, as in the case of an embroidered tie-dye, or a collage of hand-woven materials.

Liquid colour. Included in the title 'Liquid colour' are any water- or oil-based paints, inks, or dyes. There is a strong connection with the section on dyes in the textile area of the network. Dyes can be used as a form of paint, or used to dye absorbent paper, as well as conventionally on textiles.

Early work with liquids, such as paint, may not involve the use of any tools apart from the hands. The tactile information received directly by the fingers and palms increases the understanding of the medium. Once some of the properties of liquid media have been explored, further activities can be introduced to emphasise particular qualities of the paint. A variety of backgrounds cause the children to use liquids in different ways. Work with tools such as the brush can be taught here. Accidental techniques where some element of control is imposed on dribbled or spattered paint are useful for the mixes of colours which can be contrived. Different kinds of liquid can be used to extend these early skills, although the practical problems of using emulsion and oil paints with young children may cause some of these activities to be delayed.

Ordinary water-based liquids have many applications in school; the variety of printing techniques in conjunction with other media being only one development. Some printing methods would require work to be carried out in other sections of the network for successful results. Thus a pupil printing with a wood block may need to know how to cut the block, as well as the information about the liquid used.

As with solid colour a knowledge of colour, and an increase in visual perception and physical co-ordination develop alongside the practical techniques. The techniques themselves should promote this development. Specialised tools like the roller, palette-knife, spray-gun and template tend to be used later in the primary school, when some of the concepts mentioned already have been established. The network diagram therefore puts these advanced skills on the right.

Some of the aspects of colour representation may be absorbed from the activities using solid media and then applied to liquids, or vice versa. Either way some of the similarities and differences between solid and liquid media can be experienced, especially if work is undertaken in both sections at the same time.

It can be seen that the network comprises eight major sections concerning materials which behave in a similar manner. Within each section a number of activities have been arranged from left to right, so that the simplest and safest skills are encountered before the more demanding ones. Thus the pupil undertaking the work in these basic materials, and learning the various skills connected with them, is faced with a sequence of activities leading to further creative work. This is true at whatever stage a student finds himself, and would certainly apply if all the activities were covered. By using the wealth of experience gained, a pupil should be at least able to attempt original work.

Although the network has been divided horizontally into divisions it is not desirable or possible, for the techniques to be kept in separate compart-

ments. Many activities are similar for each section. This is particularly true of skills such as carving and printing which employ a whole range of materials and skills. The construction of mosaics, puppets and mobiles also come into this category. Particular connections between media can be exploited as and when required and are extremely useful if a 'project' approach is being used. Therefore if mosaics were being explored by a group of children with sufficient ability, the various links in the network would provide a choice of materials suitable for preparatory work, and the bibliography would suggest more detailed information about the more complicated activities.

Record cards would be kept as for other subjects. Here the record card would also give the teacher some idea of the kinds of materials and tools required during a forthcoming school year, and would help in planning the distribution of specialised equipment. Any money saved by this advanced planning would then be available to provide new equipment and tools for the areas of work in the network which are new to the school.

A further use for the record card in advanced planning, in conjunction with the network and bibliography, is to indicate to the non-specialist art teacher the reading he needs to do.

The network is a simple framework on which a teacher, or group of teachers, can base their art and craft teaching. On this basis it is intended that each pupil will gain experience in a diverse range of materials, and also learn how to compare and combine the properties of all of them.

No child of course should be made to work slavishly through all the techniques in network order at the expense of enjoyment and spontaneity. The network should be used judiciously after careful consideration of the needs of the pupils, and the facilities which exist in the school.

If a more systematic approach to art and craft teaching is wanted this network may serve as a useful framework, or at least a model for such a framework.

Sources

The idea of designing an art and craft network first arose when several colleagues expressed a need for information about techniques to extend their pupils further with a particular material. At the same time network analysis was being used for other subjects in the school for curriculum planning as explained in chapter 3.

Many books dealing with art and craft were examined and a large number were found to contain information about individual materials, as in the *Creative Play* series published by Batsford. Other books covered a whole range of techniques in a large variety of materials. Using a flow chart and then a network analysis it was possible to arrange many of the creative techniques into broad headings, which were then put in a logical sequence. Then book references were added.

Some authors are obviously worried that a child's creativity could be damaged by too much teacher control. At the moment it seems that only the 'experimental' approach to art and craft has been tried, whereas every attempt should be made to try out alternatives. Perhaps by arranging materials and techniques in logical hierarchies some children who do not benefit from more informal approaches will gain.

The teacher guides and stimulates the work which takes place, and provides the materials, which in themselves may stimulate the child. The teacher arranges tools and materials so that they are easy to use and are likely to attract the attention of a child wondering what to do next. The teacher must organise the equipment and its care so that creative work can take place smoothly and so that he is free to discuss, advise and guide. The network is intended to help the non-specialist teacher to do all these things. It is in the belief that teachers can and should stimulate artistic awareness with every means at their disposal that the art and craft network is presented.

References

When the network of art and craft was used in a school a list of over seventy books on various aspects of art was drawn up and numbered. These numbers can be seen on the network alongside the name of the activity. It may help some teachers to enclose a list of a few of these references which are most used. This could then form the basis of a reference library in a staffroom. Any teacher who wishes to guide pupils to a new activity, or to make suggestions from which pupils could select their own activity, can look at the network, and if the activity is not familiar, look in the books listed for ideas.

Any group of teachers can use the network and insert their own reference numbers alongside the activities as they build up their own library of reference books. This library can be built up quite quickly with the help of a librarian, or the local teacher's centre warden if there is no art and craft specialist in the school.

Since ideas change and new books appear regularly while others go out of print, each school must try to draw up its own library. As time goes on some books will fall out of use and new ones will need to be included. Some of the materials in the network may no longer be available, and others may need to be included. Here only a few books are listed and numbered so that teachers can see how this numbering system can be added to the network.

REFERENCE FOR ART AND CRAFT NETWORK ANALYSIS

1 *Creative Clay Craft*, E. Rottger. Batsford, 1962

3 *The Beginners Book of Pottery* (Part 1), H. Powell. Blandford, 1967

5 *Making Mosaics*, John Berry. Studio Vista, 1966

11 *Creative Metal Craft*, E. Rottger. Batsford, 1969

18 *Creative Textile Craft, Thread and Fabric*, Rolf Hartnung. Batsford, 1969

23 *Tie and Dye*, Anne Maile. Mills & Boon, 1965

25 *Creative Paper Craft*, E. Rottger. Batsford, 1967

28 *I can do it* (Book 3), H. Mell and E. Fisher. Schofield & Sims, 1969

30 *Pictures with Paints*, Lothar Kampman. Batsford, 1969

32 *Creative Drawing: Point and Line*, E. Rottger and D. Klante. Batsford, 1969

50 *Simple Print Making*, G. Kent and M. Cooper. Studio Vista, 1966

53 *Craft Work for Juniors*, A. Mayland. Evans, 1966

54 *Colour Crafts, Wire, Wood and Cork*, Vol. 4. Macdonald, 1971

55 *Make it yourself*, H. Danby. Heineman, 1974

56 *How to make decorations*, M. K. Skinner. Studio Vista, 1974

57 *Carton Craft*, R. Slade. Faber, 1972

58 *How to Start Carving*, C. Graveney. Studio Vista, 1974

59 *Dolls and Puppets*, M. Crockett. David & Charles, 1974

60 *How to make flying things*, Michael Bond. Studio Vista, 1975

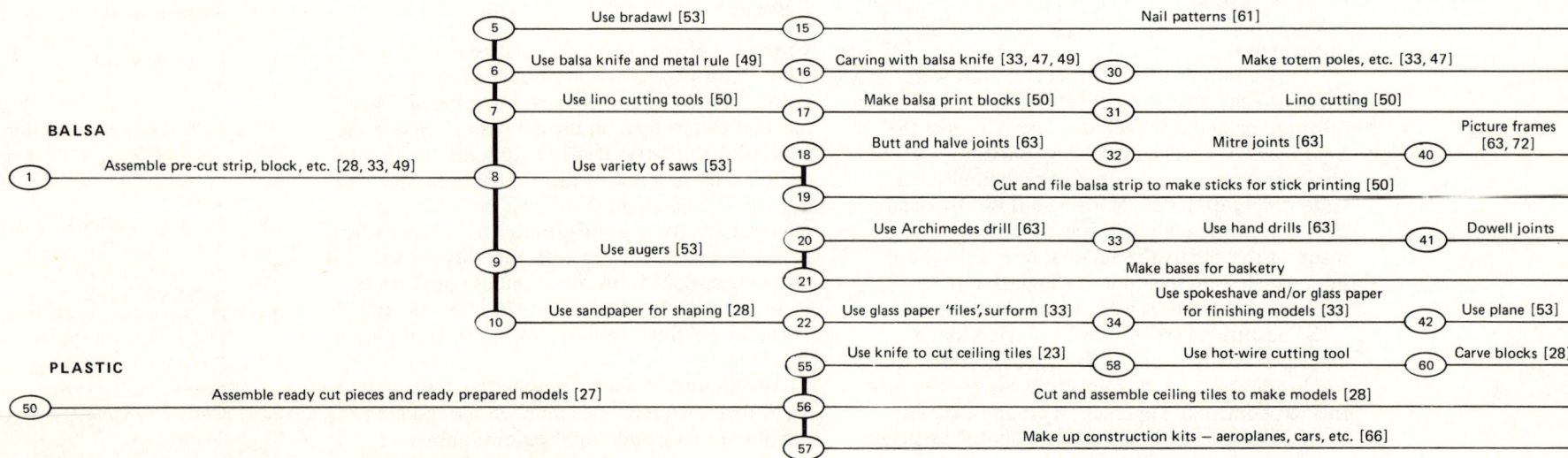

Fig. 40. *A modified network analysis of early Art and Craft curriculum.*

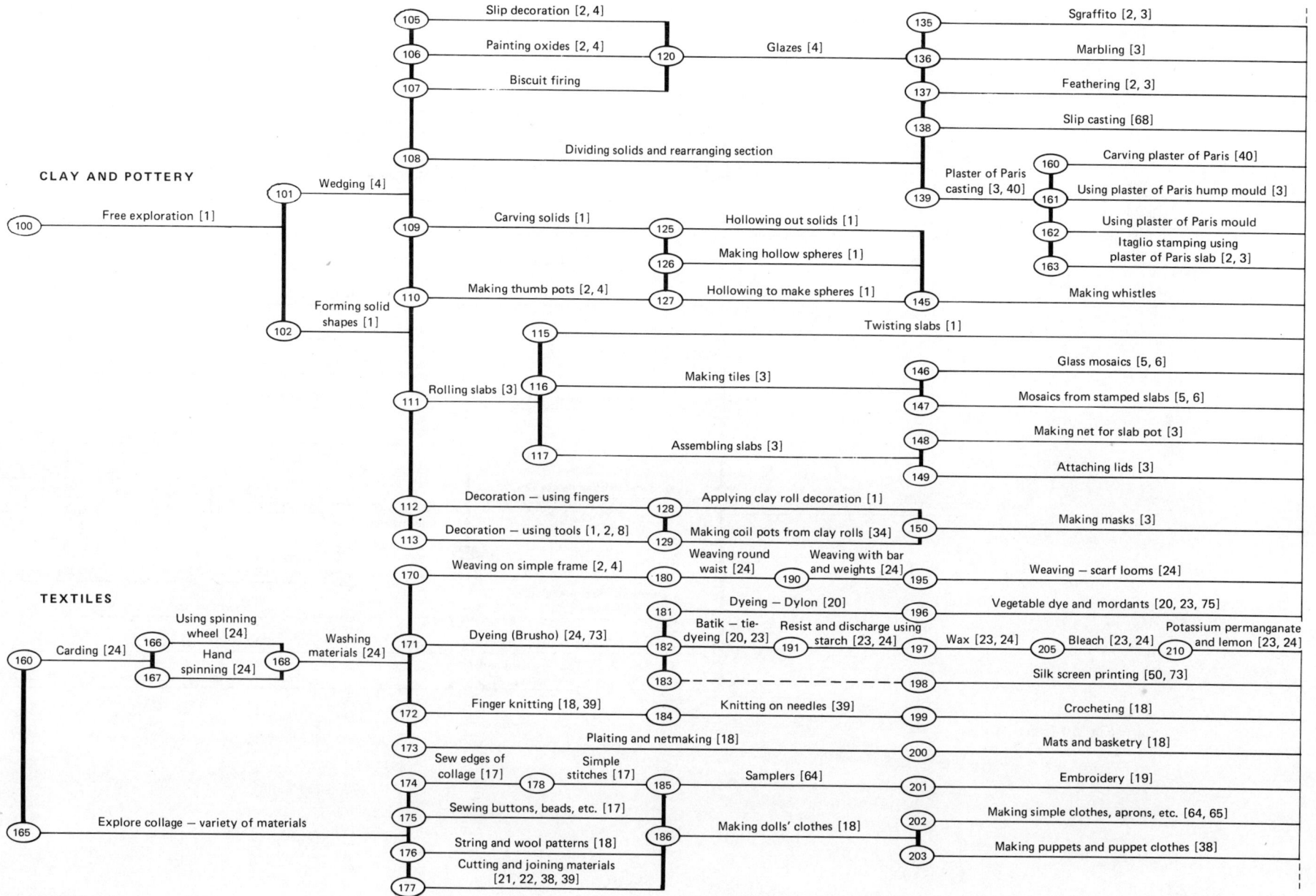

CLAY AND POTTERY

(105)	Slip decoration [2, 4]		(135) Sgraffito [2, 3]
(106)	Painting oxides [2, 4]	(120) Glazes [4]	(136) Marbling [3]
(107)	Biscuit firing		(137) Feathering [2, 3]
			(138) Slip casting [68]
(108)	Dividing solids and rearranging section		(160) Carving plaster of Paris [40]
(101) Wedging [4]		(139) Plaster of Paris casting [3, 40]	(161) Using plaster of Paris hump mould [3]
(100) Free exploration [1]	(109) Carving solids [1]	(125) Hollowing out solids [1]	(162) Using plaster of Paris mould
		(126) Making hollow spheres [1]	(163) Itaglio stamping using plaster of Paris slab [2, 3]
(110) Making thumb pots [2, 4]		(127) Hollowing to make spheres [1]	(145) Making whistles
(102) Forming solid shapes [1]	(115)		Twisting slabs [1]
(111) Rolling slabs [3]	(116) Making tiles [3]		(146) Glass mosaics [5, 6]
			(147) Mosaics from stamped slabs [5, 6]
	(117) Assembling slabs [3]		(148) Making net for slab pot [3]
			(149) Attaching lids [3]
(112) Decoration — using fingers	(128) Applying clay roll decoration [1]		(150) Making masks [3]
(113) Decoration — using tools [1, 2, 8]	(129) Making coil pots from clay rolls [34]		

TEXTILES

(170) Weaving on simple frame [2, 4]	(180) Weaving round waist [24]	(190) Weaving with bar and weights [24]	(195) Weaving — scarf looms [24]			
(166) Using spinning wheel [24]	(171) Dyeing (Brusho) [24, 73]	(181) Dyeing — Dylon [20]	(196) Vegetable dye and mordants [20, 23, 75]			
(160) Carding [24]		(182) Batik — tie-dyeing [20, 23]	(191) Resist and discharge using starch [23, 24]	(197) Wax [23, 24]	(205) Bleach [23, 24]	(210) Potassium permanganate and lemon [23, 24]
(167) Hand spinning [24]	(168) Washing materials [24]	(183) ---	(198) Silk screen printing [50, 73]			
(172) Finger knitting [18, 39]	(184) Knitting on needles [39]	(199) Crocheting [18]				
(173) Plaiting and netmaking [18]	(200) Mats and basketry [18]					
(174) Sew edges of collage [17]	(178) Simple stitches [17]	(185) Samplers [64]	(201) Embroidery [19]			
(165) Explore collage — variety of materials	(175) Sewing buttons, beads, etc. [17]	(202) Making simple clothes, aprons, etc. [64, 65]				
(176) String and wool patterns [18]	(186) Making dolls' clothes [18]	(203) Making puppets and puppet clothes [38]				
(177) Cutting and joining materials [21, 22, 38, 39]						

Fig. 40 (part 2).

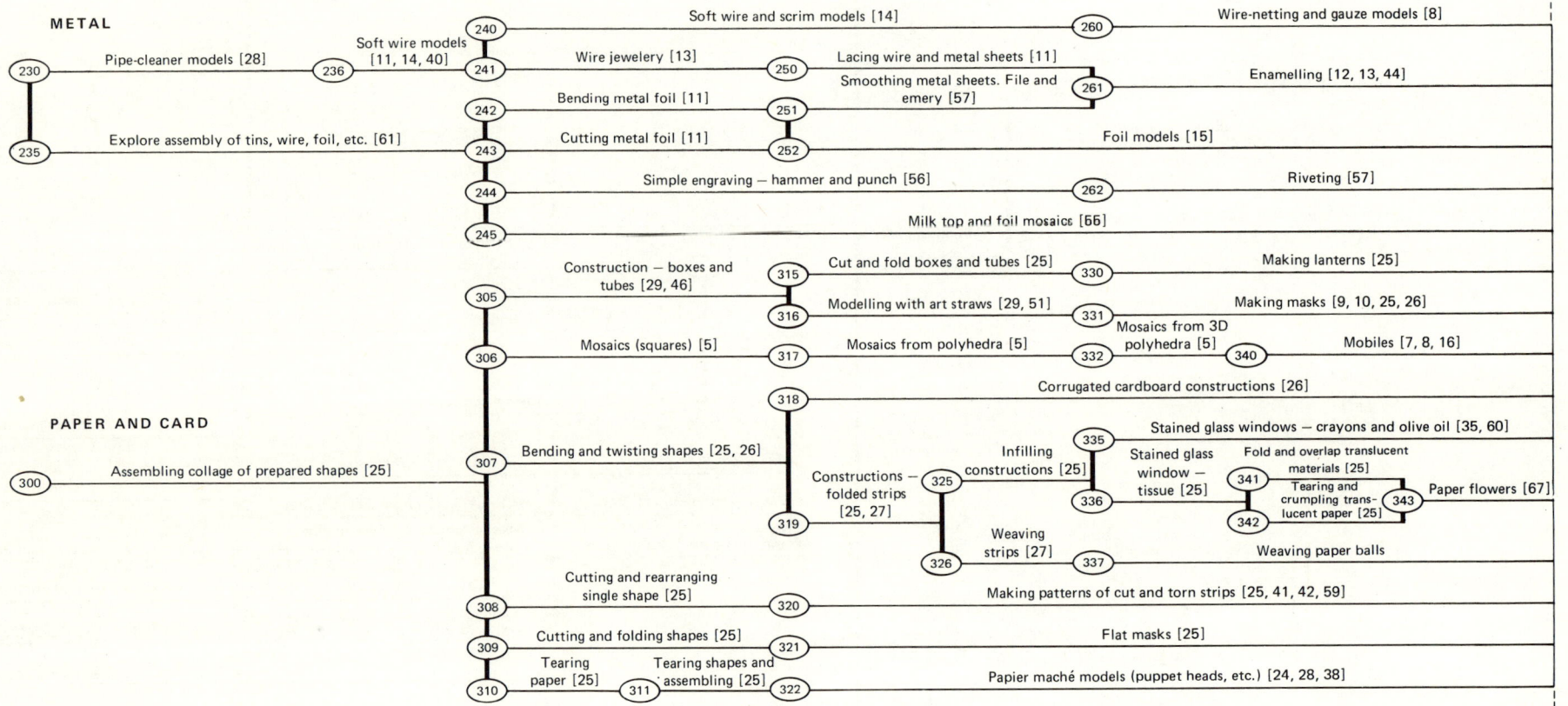

METAL

Soft wire and scrim models [14]

Wire-netting and gauze models [8]

(240) (260)

Pipe-cleaner models [28]

Soft wire models [11, 14, 40]

Wire jewelery [13]

Lacing wire and metal sheets [11]

Enamelling [12, 13, 44]

(230) (236) (241) (250) (261)

Bending metal foil [11]

Smoothing metal sheets. File and emery [57]

(242) (251)

Explore assembly of tins, wire, foil, etc. [61]

Cutting metal foil [11]

Foil models [15]

(235) (243) (252)

Simple engraving — hammer and punch [56]

Riveting [57]

(244) (262)

Milk top and foil mosaics [56]

(245)

Construction — boxes and tubes [29, 46]

Cut and fold boxes and tubes [25]

Making lanterns [25]

(305) (315) (330)

Modelling with art straws [29, 51]

Making masks [9, 10, 25, 26]

(316) (331)

Mosaics (squares) [5]

Mosaics from polyhedra [5]

Mosaics from 3D polyhedra [5]

Mobiles [7, 8, 16]

(306) (317) (332) (340)

Corrugated cardboard constructions [26]

(318)

PAPER AND CARD

Stained glass windows — crayons and olive oil [35, 60]

Bending and twisting shapes [25, 26]

(335)

Infilling constructions [25]

Stained glass window — tissue [25]

Fold and overlap translucent materials [25]

Assembling collage of prepared shapes [25]

(300) (307)

Constructions — folded strips [25, 27]

(325) (336) (341)

Tearing and crumpling translucent paper [25]

(343) Paper flowers [67]

(319) (342)

Weaving strips [27]

(326)

Weaving paper balls

(337)

Cutting and rearranging single shape [25]

Making patterns of cut and torn strips [25, 41, 42, 59]

(308) (320)

Cutting and folding shapes [25]

Flat masks [25]

(309) (321)

Tearing paper [25]

Tearing shapes and assembling [25]

Papier maché models (puppet heads, etc.) [24, 28, 38]

(310) (311) (322)

Fig. 40 (part 3).

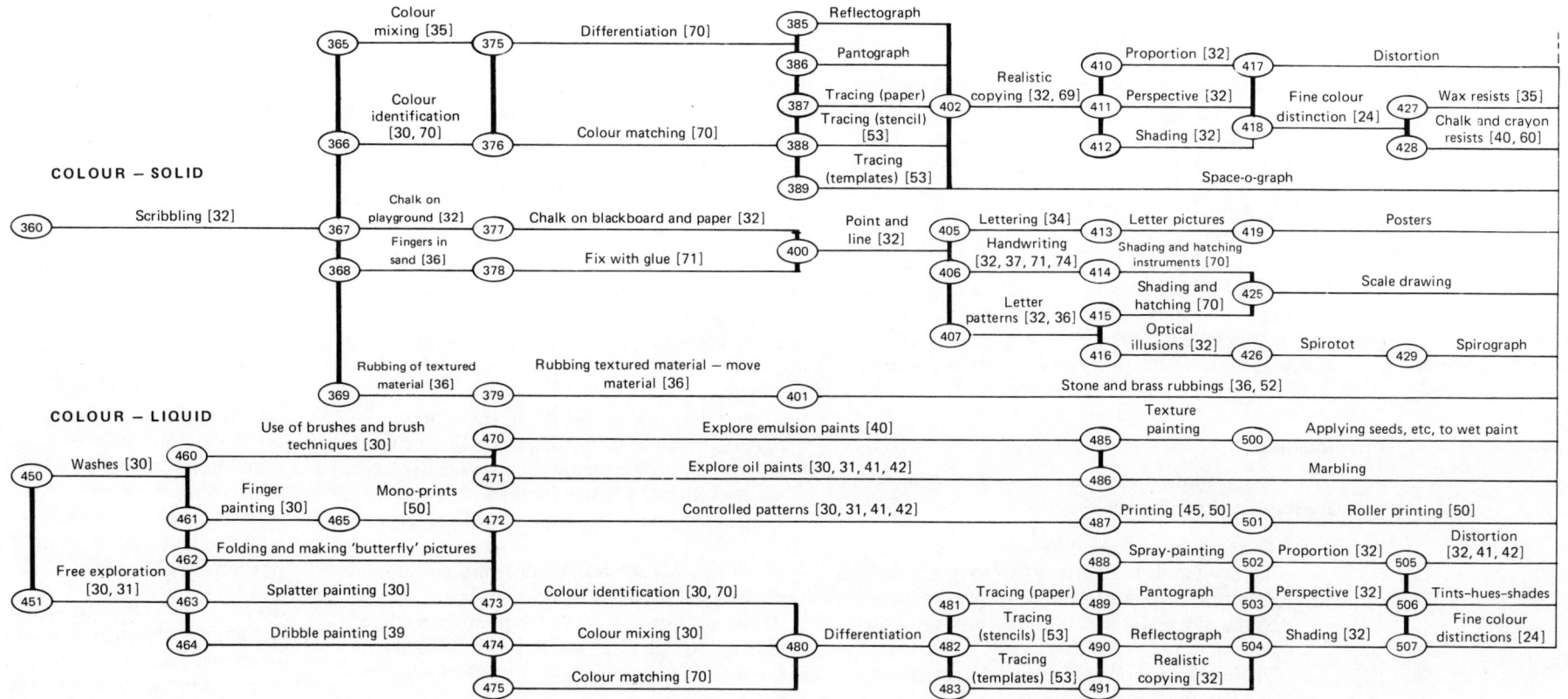

Fig. 40 (part 4).

6 · English, geography and history

In subjects like mathematics and science there is a fairly easily recognisable structure. Once this basic structure has been identified and choices have been made the various activities involved can be sorted into a logical system.

This chapter contains first design studies for similar treatment of subjects for which it might seem less suitable. As these are first design studies there is undoubtedly scope for alteration and modification before they can be effectively used. But they are included to demonstrate that the technique outlined in the previous chapters can equally well be applied to humanities subjects.

Reading

There is a need for structure in the teaching of reading for the following reasons:

1 There is such a large turnover of teachers in infant schools that without a recognisable structure and properly kept records the work of the individual pupil will be constantly disrupted.
2 There are rules of spelling and grammar which can be generally applied.
3 For discovery methods to be of value a pupil must first be able to read reference books, and know how to search for the right ones. The necessary reading skill must therefore be accompanied by other skills, sometimes known as 'library' skills.
4 Generally it is recognised that the greater the freedom to follow an individual course of learning, the more need there is for an underlying structure.

While it may be difficult to agree on a logical progression in the teaching and learning of reading skills, it seemed worth producing a design study for critical examination. It is to be emphasised that this is simply an attempt to show how network analysis might be used to form the basis of an integrated scheme for the teaching of reading and English; the design has not been tried out. Whatever plan is used, and however a pupil is taught, success will ultimately depend on an orderly succession of steps and good record-keeping. Network analysis can contribute to this. The aim in designing the English network was not to produce a network in such detail that course designers could use it for the production of teaching materials. A much more detailed network would be required for this purpose. The aim was to look at the possibility of producing an overall structure.

Figure 41 is a network of the simple discriminations involved in the recognition of letter shapes.

If the activities on the network are considered in relation to the letter shapes it can be seen that it is only after a pupil knows the difference between:

large and small 5
round 3
half round 6
straight and crooked 4
upright 7
long and short 8

that he will be able to grasp fully the shapes of the letters, S, l, i and s.

The network analysis of infant–primary mathematics was found of value in allocating the existing teaching materials in the school, and could be used as an aid to the design and allocation of teaching materials.

When a pupil cannot carry out an activity and a decision is made to design remedial teaching materials, a lot of care and thought goes into the production of such material. An effort is made to try to analyse what sort of pre-knowledge is required. In the case of letter recognition, an analysis of the simple discriminations involved would be made before the teaching material is designed and tested. The simple discriminations are arranged in logical sequence. The dependence of one activity upon another can be seen clearly, and so if the pupil cannot recognise a letter shape, the sequence of activities required can be identified.

If the network analysis is used as a basis, this might be an order in which the simple discriminations involved in letter recognition could be taught.

Stage 1. Activities with teaching material to enable the pupil to distinguish round shapes, and the differences between straight and curved or straight and crooked, and large and small. When this teaching has been given then upper case and lower case (small and capital letters) 'O' could be introduced.

Stage 2. Teaching materials to teach – half round, upright, long and short. Letters introduced – lower case: l, i, s; upper case: S.

Stage 3. Teaching materials – the distinction between up and down. The pupil should be able to count up to three. Upper case letters: B, D, U, L; lower case letters: a, m, n, r, u.

Stage 4. Teaching materials — ability to distinguish diagonal. Upper case letters: none; lower case letters: b, d, f, g, h, j, p, q, t.

Stage 5. Teaching materials — position on clock face. Upper case: A, K, M, N, R, V, W, X, Y, Z; lower case: k, v, w, x, y, z.

Stage 6. Upper case: C, G, Q; lower case: c, e.

An examination of the network will show which teaching materials are required, and then which letters can be introduced.

The first activity before entering any teaching scheme would be an initial test to ascertain the correct starting point for that individual pupil. Once this has been done it is easy to decide at which point in a network based scheme the pupil should start. It is not envisaged that all pupils will start with activity 1 and then proceed right through the whole programme of work.

Since the network analysis of the teaching of reading/English would be numbered in sequence

it is a simple matter to base a record card on the network and to record the progress of the pupil. It is essential for any scheme of work if it is to be effective, that adequate records are kept. It is by the study of the results of tests that the overall programme of work can be evaluated and progressively improved.

Having looked in detail at a small section of the reading network we can now turn to a section of the whole English network and examine it more closely (figure 42).

Because of the way the technique works, it might appear that there is a division of activities into watertight compartments. Since there are three main streams of discrimination involved: visual, auditory, and motor, it must not be thought that they are dealt with in isolation.

From the drawing of the network this is bound to appear so. The lines of progression of visual, auditory, and motor skills are dealt with separately. It is when this has been done that the dependences of one upon another can be examined. It

then becomes obvious, for example, that the visual discrimination of uprightness (activity number 126) should be linked with the activity, drawing lines vertically (activity number 139). Once the separate lines of progression are decided in a logical sequence, the dependence of one line upon another can be sorted out. It is quite an easy matter to move activities along horizontally in order to make allowance for these dependences. As this network has by no means reached the final stage of use in school, it is highly likely that there are a number of inconsistencies and errors. It must be emphasised that this is a design study to show how the technique may be used. If it does this, it will have served its purpose. The preliminary listing and ordering of the basic progression of skills and activities is an extremely complex task and would involve a great deal of research, assessment and modification.

In the later stages of the network are lists of words which pupils ought to be taught. This is not to suggest that they should be given spelling lists,

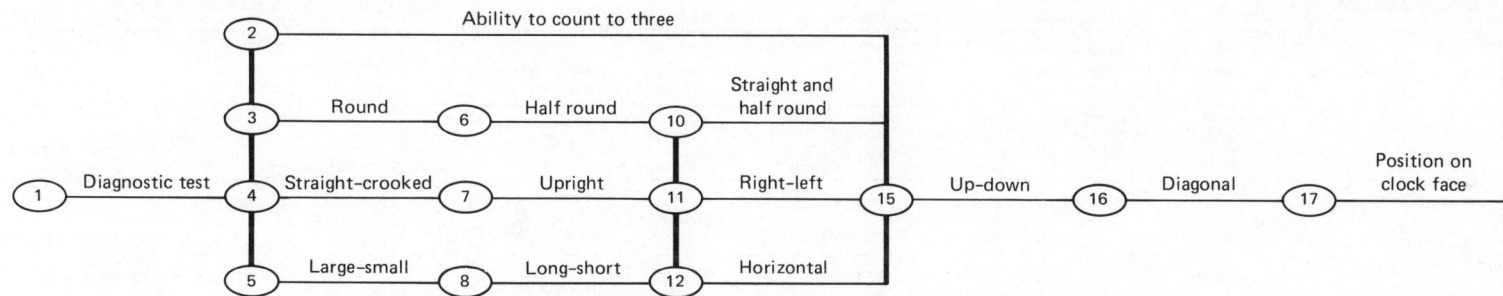

TYPE OF LETTER	STAGE 1	STAGE 2	STAGE 3	STAGE 4	STAGE 5	STAGE 6
Upper case capital	O	S	B D U L		A K M N R V W X Y Z	C G Q
Lower case small letters	o	l i s	a m n r u	b d f g h j p q t	k v w x y z	c e

Fig. 41. *A modified network analysis of early discriminations involved in letter recognition.*

VISUAL DISCRIMINATION

'Look and Say' – *Queensway – Key Words – Time for Reading*, etc.

Choice of reading scheme(s) e.g.

'Phonic' – *Royal Road – Fun with Phonics – Stott*, etc.

'Different Media' – *Words in colour – I.T.A. – Colour Story Reading*, etc.

Round — Half round — Straight and half round

Straight-curved crooked — Upright — Right-left — Up-down — Diagonal — Position on clock face

Large-small — Long-short — Horizontal

Letters:
o — l i s — a m n r u — b d f g h j p q t — k v w x y z — c e

Sorting pictures into story sequence — Sorting pictures into story sequence — Reading picture story – anticipate end — Caption book and tape recordings — Start picture–word book

Picture stories

Matching – shapes — Jigsaws of shapes — Matching letters learnt — Matching letters learnt — Matching letters learnt — Matching letters learnt — Matching letters learnt

Essential social vocabulary, e.g. Moneyhull

AUDITORY DISCRIMINATION

Matching – pictures — Sorting shapes into sets — Sorting letters into sets — *Choice of individual reading scheme (S.R.A., Stott, Ward-Lock Workshop, etc.)*

Teacher reading story, tape recordings, records — Teacher reading story, tape recordings, records — Teacher reading story, tape recordings, records — Teacher reading story and answering questions about pictures — *Letter-sound links*
Making up stories about pictures — Making up stories about pictures — *Initial consonants* — Telling complete story to children

Possible progression

Tongue twisters — Initial vowels
a < AT / APE — Consonant before vowel < BAG / BAY — Vowel before consonant < ABOUT / ABLE — o < POT / NOW — a < PAT / HAY — f h j — v w k z — *Silent letters:* w.b.g.t.p.l.k.m gh.

Nursery rhymes — i < it / item — < BITE / BIT — < ITALY / ISLAND — l – RIDDLE — u < UNDER / IMPUTE — *Double consonants* bb dd ch ff sh gg — ck wh

Songs with choruses — e < EGG / EGO — < BEG / BE — < EGG / EVIL — i < PIN / SITE — *Initial consonant digraphs* th < THUMB / THUS — qu — sk — y < YES BABY FLY — sc cr cl

Percussion games — o < otter / owe — < POT / NOW — < ONSET / OVER — *Initial consonant blends* sm sn sl — fr- pr- gr- tr- bl- fl- st- sp- br- dr- gl — sw — pl — sc sk ue

Tapes of sounds: practical recognition — u < UGLY / USE — < PUG / TRUE — < UGLY / UNIT — s-(was)=z — *Vowel digraphs* ai oo < FOOD / WOOD — oa ou < COULD / FOUND — < COW / SNOW — oy ie < pie / field ee ea < HEAD / BEAD ei

Vowel consonant blends ar- car ur- murder ir- girl or- for — g (eiy) = J — c (eiy) = S er – HER

Phonic elements

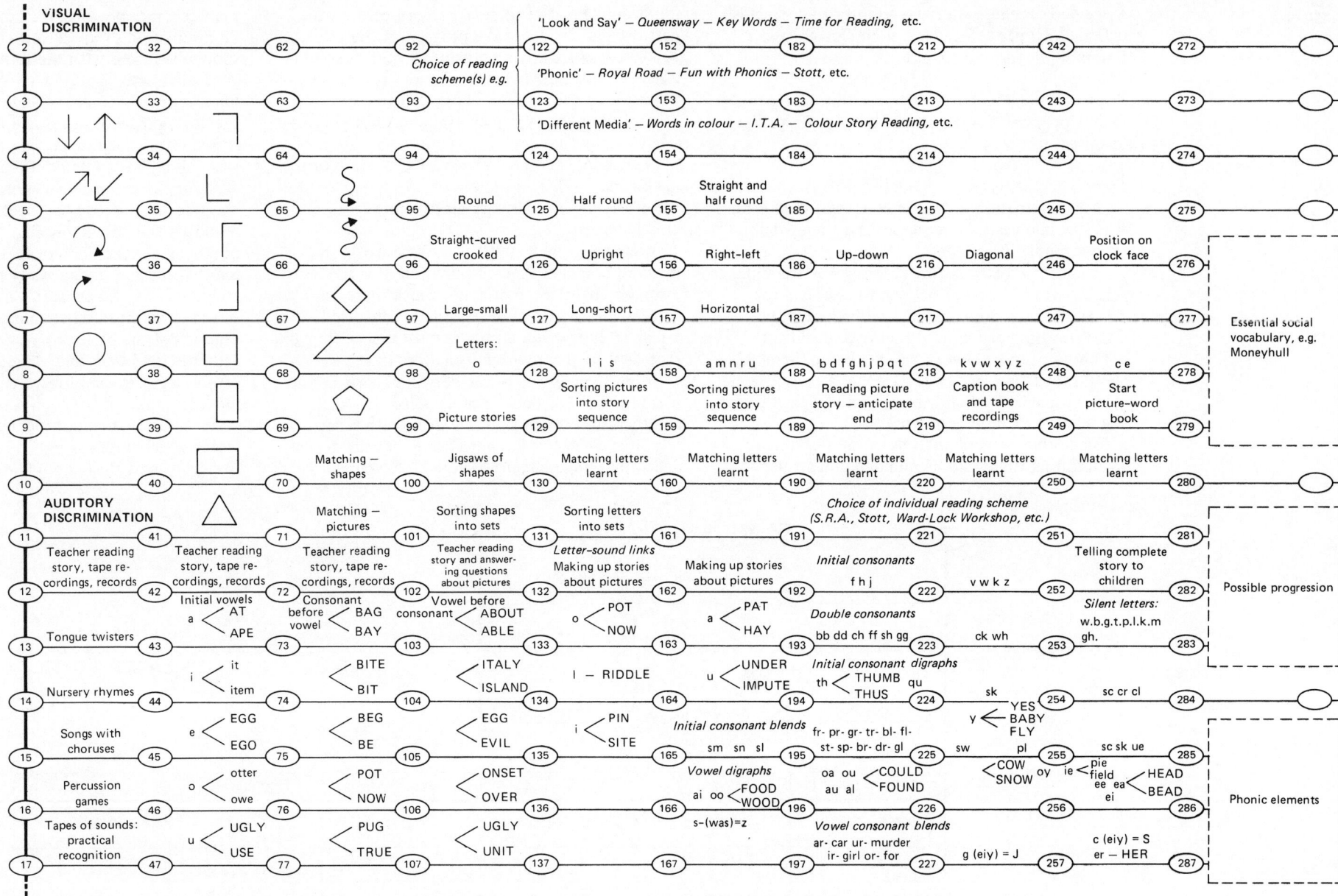

Fig. 42. *A section of a modified network analysis of reading and early English curriculum.*

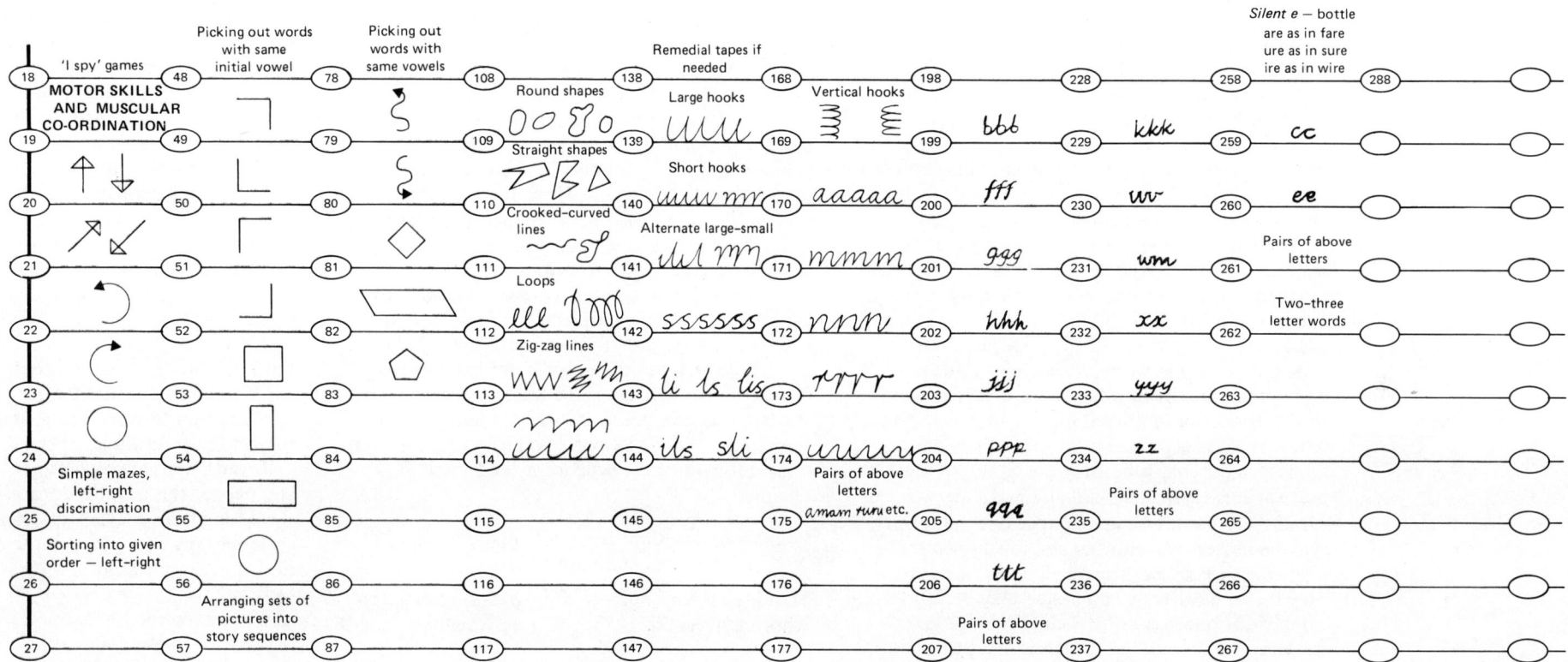

Fig. 42 (part 2).

Text labels within the figure:

- 'I spy' games
- MOTOR SKILLS AND MUSCULAR CO-ORDINATION
- Picking out words with same initial vowel
- Picking out words with same vowels
- Round shapes
- Straight shapes
- Crooked-curved lines
- Loops
- Zig-zag lines
- Remedial tapes if needed
- Large hooks
- Short hooks
- Alternate large-small
- Vertical hooks
- Pairs of above letters
- Silent e — bottle / are as in fare / ure as in sure / ire as in wire
- Pairs of above letters
- Two-three letter words
- Simple mazes, left-right discrimination
- Sorting into given order — left-right
- Arranging sets of pictures into story sequences
- Pairs of above letters
- Pairs of above letters
- Pairs of above letters

for this is a sterile approach which often fails. It is far better that pupils learn to spell those words which they wish to use in their creative work.

The lists of words are included for specific purposes. In the case of the phonic list, there may be value in bringing to the attention of the pupils those words which embody certain sounds which can be easily identified, and can be built up into a vocabulary where the sound and spelling are linked. The other list is of words which are necessary for children to learn if they are to function efficiently in our society. Such words as 'poison, danger, keep out, beware, etc.' quite clearly have a value, and should be taught as a deliberate policy.

There is a section in which it is suggested that a reading scheme or schemes should be chosen. This is a choice which will have been taken by the teacher, and will probably be a part of the school policy. At a time when all the pupils in a class were streamed and were proceeding through a common course of learning this sort of choice might have been of critical importance. However, when a course is prepared for individuals the choice is not what is best for the class, but what is best for this particular child at a particular stage of his development. For some pupils one approach may be better than another and it is hoped that an analysis of reading skills will aid this choice. It may be found that one approach may be very effective at one stage of the child's development but may fail at another. Certainly if a pupil fails to make progress, another approach should be tried. One would suggest therefore that no one method of teaching reading should be adopted to the exclusion of others, but that all should be used when they are shown to be required by the individual pupil. There is a section which gives some examples of the sort of exercises which could be used. This is only inserted as an example of one approach and is by no means complete.

The latter end of the network (not illustrated) includes some activities which might rightly appear in a middle school or comprehensive network, as for example the methods which pupils should be taught in order to find information from reference books in open-ended exploration, and methods of study which are most effective. However, in many primary schools pupils are encouraged to engage in research of this type in their various projects. If they are to gain any benefit from this exploratory activity, the techniques of research should be taught. In the same way the various types of reading should be taught. Reading aloud is not the only aim for pupils. Pupils need to be taught how to skim read, how to sort out the relevant from a mass of detail, and so on. These techniques of reading should be broken down into stages and introduced to pupils so that they may use their reading skills as tools for various purposes.

It is hoped that enough has been said now to enable anyone who considers that there is value in this sort of approach, to take this rather crude design and refine and modify it until it becomes an instrument which may be used with effect. In the teaching of reading and early language skills the value of a good record-keeping system, such as this treatment of the syllabus makes possible, is obvious.

Possible progression: from network numbers 281, 282, 283 (see Appendix, p. 118).

Phonic elements: suggested progression from network numbers 285, 286, 287 (see Appendix, p. 118).

Geography

In many schools for younger pupils geography and history are linked into what is called 'social studies'. There may be a danger in this sort of loose association, in that since the aims may not be clearly defined, no one can judge what any one child knows, nor what any group knows. There is the danger that essential knowledge may not be taught and what is worse, no one may be aware that this is so.

It is essential that those who are to teach inter-disciplinary courses should have very clearly in mind what the boundaries are to be, and what exactly is to be included in the course. Well-qualified teachers with many years of experience may be able to carry in their minds the complexities of the subject area, but others less well-qualified, and lacking experience need every bit of help they can be given.

If first of all the exact boundaries of the separate subjects are defined, and the subject content listed, surely it will be easier to decide what topics will be included in the interdisciplinary course. The various dependences of one aspect of the work upon another can be seen and full advantage taken of common ground.

In geography there have been large changes of content and approach in higher education, and this is percolating into schools for older pupils. The new emphasis is on spatial and social interaction. Data is collected and quantitative predictions made, based upon theoretical mathematical models. This is in addition to work in the areas into which the subject has traditionally been divided.

If our pupils are going to be able to understand any part of this advanced learning, we should keep these long-term goals in geography in mind and consider how best to teach the early discriminations so that concepts are formed which can be used later on.

If we keep the hierarchical model of the learning process in mind and apply this to learning in geography we will need to give the pupils experiences of the gross differences between the various aspects of geography, and gradually refine these until concepts of land masses, mountains, climate, vegetation types, etc. have been acquired.

Broadly speaking there are two main ways of setting about this in the primary school. The teacher can either take one aspect of geography, for example, transport, and study this first in this country and then expand this into a world-wide study, or approach each country, or continent in turn. There are valid arguments to be advanced for either approach. In any case the teacher will be concerned with the presentation of differences, and similarities. During the pupil's time in the school these discriminations will gradually be refined. A network can be drawn which is divided into lines of activities based on topic areas such as the people who live in a country, the cities, the mountains or hills, the rivers, lakes, transport, houses, raw materials and industry, climate, and crops. The network is divided vertically into columns, one for each land area. Since most geography would be about the pupils' own environment, Europe and the United Kingdom are repeated at intervals in this example.

In addition to the formal periods of teaching about geography there would be a continuous study of the various aspects of weather. In the early stages this will merely consist of a statement, that it is sunny and warm or rainy, snow, etc. In the later stages the pupils will measure and record the temperature and rainfall. The connection between the clouds and weather will be noted. The connection between the school records and the weather forecasts on radio and television will be drawn,

and all this will be linked with the other forms of measurement in mathematics.

Most schools plan and organise visits of various types and some of these will be visits to places of geographical interest within reach of the school. During each of these visits the children's attention will be drawn to appropriate details such as vegetation, crops, land form, housing materials, and farm animals.

Full use will be made of maps and plans and the pupils' conception of the connection between the symbols on maps and the objects represented will be refined. There are some doubts as to how far maps are of value in the primary school. Care should be taken to structure the presentation of maps. In the early stages plans of the classroom should be introduced, and only when the connection between the representation and what it represents is firmly established should the more refined map-making conventions be used.

A section of a network design is shown overleaf. An effective record card and record-keeping system can easily be based on such a network (figure 43).

	ASIA	AFRICA	AMERICA	EUROPE	U.K.	ASIA	AFRICA	AMERICA	EUROPE
COUNTRIES (1)	China (21)	South Africa (41)	South America (61)	Scandinavia (81)	(101)	Australia, East Indies (121)	West Africa (141)	Central America (161)	Eastern Europe (181)
People (2)	Chinese and Mongols (22)		Indian, Spanish, Portuguese (62)	(82)	London, Cardiff, Edinburgh, Birmingham, Manchester, Glasgow (102)	Sydney, Melbourne, Singapore, Manila, Jakarta (122)	(142)	(162)	(182)
Cities (3)	Peking, Canton, Shanghai (23)	Pretoria, Johannesburg, Cape Town, Durban, (43)	Buenos Aires, Santiago, Sao Paulo, Rio de Janiero (63)	Copenhagen, Stockholm, Oslo (83)	(103)	(123)	Accra, Freetown, Kano, Lagos, Kinshasa (143)	Mexico, Kingston, Panama, Havana, Caracas, Port of Spain (163)	Warsaw, Sofia, Bucharest, Instanbul, Budapest, Belgrade (183)
Mountains (4)	(24)	(44)	Brazilian Highlands, Andes (64)	(84)	Grampians, Welsh mountains, Pennines (104)	Great Dividing Range (124)	(144)	(164)	Carpathian mountains (184)
Rivers (5)	Yang Tse Kiang, Hwang Ho, Si-kiang (25)	Orange, Limpopo (45)	Rio de la Plata, Orinoco, Amazon (65)	(85)	(105)	Murray-Darling (125)	Zaire, Niger, Benue (145)	(165)	Danube, Oder, Vistula, Elbe (185)
Transport (6)	All forms, Grand Canal (26)	All forms (46)	All forms (66)	All forms (86)	All forms (106)	All forms (126)	(146)	Panama Canal (166)	(186)
Houses (7)	(27)	(47)	(67)	(87)	(107)	(127)	(147)	(167)	(187)
Raw materials Industry (8)	Oil, iron, mixed industries (28)	Mixed industries. Mining — gold, coal, iron, diamonds (48)	Copper, iron (68)	Paper, pulp, iron ore, fishing (88)	(108)	Oil, iron, copper, rubber, mixed industry (128)	Oil (148)	Oil, asphalt, bauxite (168)	Timber, oil, mixed industries, farming (188)
Climate (9)	(29)	Mediterranean, sub-tropical (49)	Mediterranean, sub-tropical, tropical (69)	Temperate, Arctic (89)	Temperate (109)	Temperate, Mediterranean, sub-tropical, tropical (129)	Tropical, sub-tropical (149)	Sub-tropical, tropical (169)	Continental temperate (189)
Crops (10)	Wheat, rice, cotton, tea (30)	Sheep, citrus fruit, wine (50)	Sheep, cattle, timber, coffee, cocoa, cotton (70)	Wood, dairy and farm produce (90)	(110)	Tropical–Mediterranean temperature fruit, meat, wool, tea, rubber, wine (130)	Oil, nuts, cocoa, cotton (150)	Bananas, sugar, coffee, cocoa (170)	Cereals, meat, Mediterranean temperature fruit (190)
Weather recording (11)	(31)	(51)	(71)	(91)	(111)	(131)	(151)	(171)	(191)
Plans and maps (12)	(32)	(52)	(72)	(92)	(112)	(132)	(152)	(172)	(192)
(13)	(33)	(53)	(73)	(93)	(113)	(133)	(153)	(173)	(193)

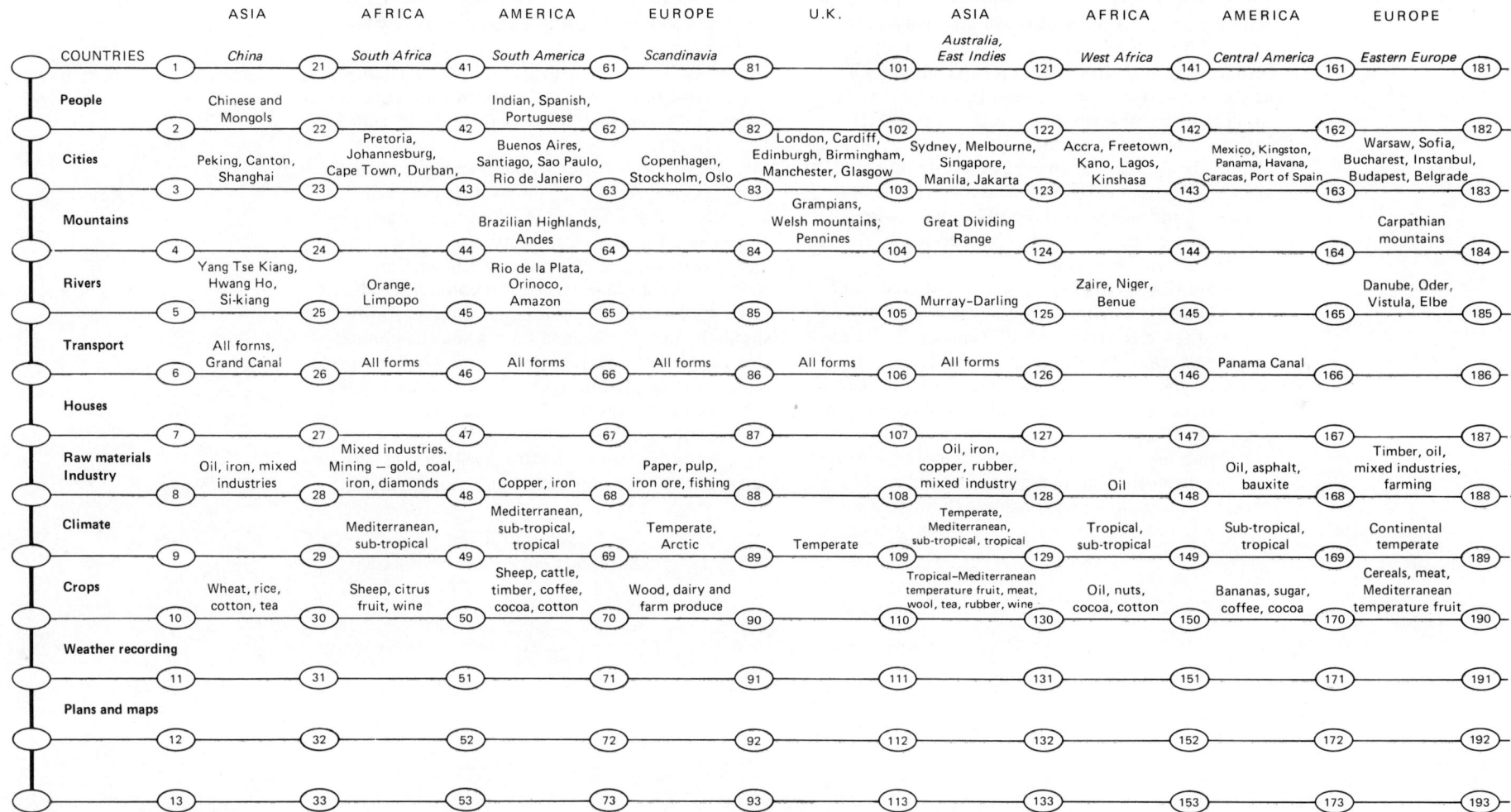

Fig. 43. *A section of a modified network analysis of the geography curriculum for young pupils.*

History

As with geography there are two main ways in which the learning of the pupils may be directed. The first is to take one aspect of the life of people who lived in this world, and follow it through the centuries. This gives the pupils a grasp of the sweep of history. The second approach is to study life at a particular time in history and then proceed to another period, and so on. The history network has been designed to make either approach possible.

Lines of progression for a British school might be

1 Kings and queens
2 Political leaders, prime ministers
3 Houses
4 Clothes
5 Land transport
6 Sea transport
7 Exploration
8 Battles, armour, weapons
9 Discoveries and scientists
10 Artists, writing
11 Painting
12 Music
13 Farming and agriculture
14 Industry
15 World events
16 Blank activities for teachers to fill in as they think fit.

The progression through time along the lines of history starts at 2500 BC and proceeds by five-hundred-year sections up to the birth of Christ. After the birth of Christ the sections are one-hundred years long, and, from 1800, fifty years.

If it is thought better there is no reason why all the sections should not be of equal size. It seemed however that there is so much more known about more recent events that it might be wiser to go into these sections in more detail. In addition children take more interest in events which impinge more directly on their own lives. In the early stages pupils will be told stories about the lives of heroes and kings and queens, and it is only gradually that they will build up a conception of the sequence of historical events.

Every opportunity should be taken to link history with other subjects. For instance the lives and work of the great mathematicians of the past provide a link with mathematics. Similarly the history of discovery can link history with geography.

There has been a tendency to avoid the rigid study of dates, and kings and queens in a formal manner. This widens the door to a broader study of history in which battles and sovereigns no longer predominate but the lives of ordinary people form an equally important part of the course. If we are to make any real balanced view of history the teaching of the progression of ideas about architecture, music, painting and the arts in general should surely have a place.

We are very conscious today of 'progress' and this is often linked with technological development. The way in which various inventions have affected the lives of people in different parts of the world, and how these have increased rapidly in the last few decades is certainly worthy of study.

The network illustrated in figure 44 is a design study of the sort of curriculum which may enable teachers to teach history more systematically, and by using a record card linked with the network, keep some sort of check on what has been taught.

If the project approach is adopted full use should be made of local resources. The local library service, for instance, may supply not only books, charts, film strips, films and records or tapes, but also artefacts. Once the overall plan of work has been agreed all these aids can be deployed to the best advantage.

Other local resources may be castles, houses, ancient sites of camps, museums, and sometimes curios from the homes of the pupils. The local historical society may be prepared to help in many ways. They may have a panel of speakers who can visit the schools to give illustrated talks. Many towns are becoming increasingly aware of their heritage and often there are collections of relics of interest. Sometimes there is a collection of documents and other items of historical interest in the county archives which may be seen.

In addition in many parts of the country old buildings like wind- or water-mills, canals, factories and other types of industrial architecture are being preserved and should be used in teaching history wherever possible.

It would be an interesting and useful exercise for any teacher to take the network and insert all the local places of historical interest which can be visited by pupils.

History

Fig. 44 (part 1). *A modified network analysis of primary history curriculum.*

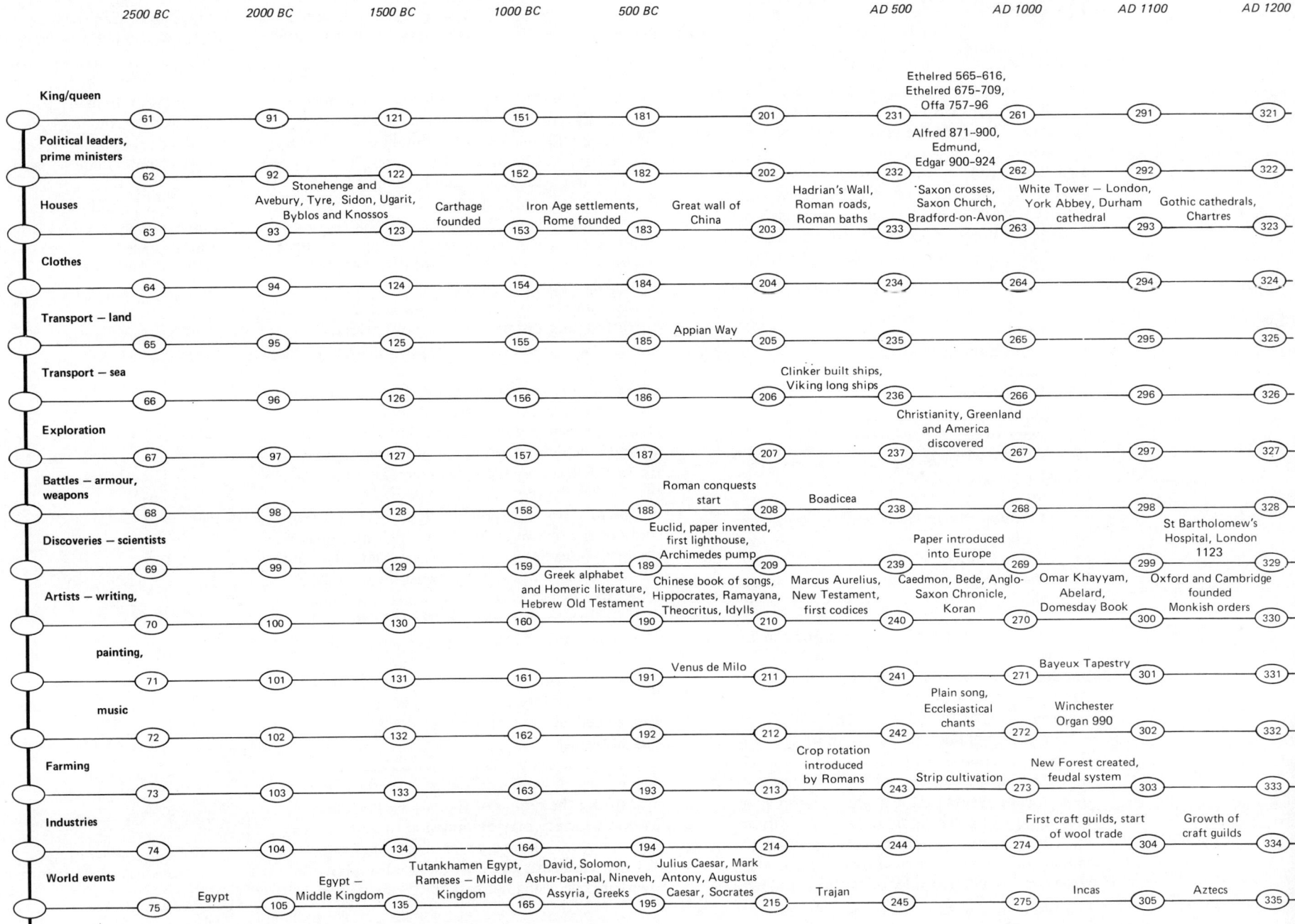

Category	2500 BC	2000 BC	1500 BC	1000 BC	500 BC	AD 500	AD 1000	AD 1100	AD 1200
King/queen	61	91	121	151	181	201	231	261	291	321
Political leaders, prime ministers	62	92	122	152	182	202	232	262	292	322
Houses	63	93	123	153	183	203	233	263	293	323
Clothes	64	94	124	154	184	204	234	264	294	324
Transport — land	65	95	125	155	185	205	235	265	295	325
Transport — sea	66	96	126	156	186	206	236	266	296	326
Exploration	67	97	127	157	187	207	237	267	297	327
Battles — armour, weapons	68	98	128	158	188	208	238	268	298	328
Discoveries — scientists	69	99	129	159	189	209	239	269	299	329
Artists — writing,	70	100	130	160	190	210	240	270	300	330
painting,	71	101	131	161	191	211	241	271	301	331
music	72	102	132	162	192	212	242	272	302	332
Farming	73	103	133	163	193	213	243	273	303	333
Industries	74	104	134	164	194	214	244	274	304	334
World events	75	105	135	165	195	215	245	275	305	335

Annotations:

- King/queen (AD 500): Ethelred 565–616, Ethelred 675–709, Offa 757–96
- Political leaders, prime ministers (AD 500): Alfred 871–900, Edmund, Edgar 900–924
- Houses: Stonehenge and Avebury, Tyre, Sidon, Ugarit, Byblos and Knossos; Carthage founded; Iron Age settlements, Rome founded; Great wall of China; Hadrian's Wall, Roman roads, Roman baths; Saxon crosses, Saxon Church, Bradford-on-Avon; White Tower — London, York Abbey, Durham cathedral; Gothic cathedrals, Chartres
- Transport — land: Appian Way
- Transport — sea: Clinker built ships, Viking long ships
- Exploration: Christianity, Greenland and America discovered
- Battles — armour, weapons: Roman conquests start; Boadicea
- Discoveries — scientists: Euclid, paper invented, first lighthouse, Archimedes pump; Paper introduced into Europe; St Bartholomew's Hospital, London 1123
- Artists — writing: Greek alphabet and Homeric literature, Hebrew Old Testament; Chinese book of songs, Hippocrates, Ramayana, Theocritus, Idylls; Marcus Aurelius, New Testament, first codices; Caedmon, Bede, Anglo-Saxon Chronicle, Koran; Omar Khayyam, Abelard, Domesday Book; Oxford and Cambridge founded Monkish orders
- painting: Venus de Milo; Bayeux Tapestry
- music: Plain song, Ecclesiastical chants; Winchester Organ 990
- Farming: Crop rotation introduced by Romans; Strip cultivation; New Forest created, feudal system
- Industries: First craft guilds, start of wool trade; Growth of craft guilds
- World events: Egypt; Egypt — Middle Kingdom; Tutankhamen Egypt, Rameses — Middle Kingdom; David, Solomon, Ashur-bani-pal, Nineveh Assyria, Greeks; Julius Caesar, Mark Antony, Augustus Caesar, Socrates; Trajan; Incas; Aztecs

74

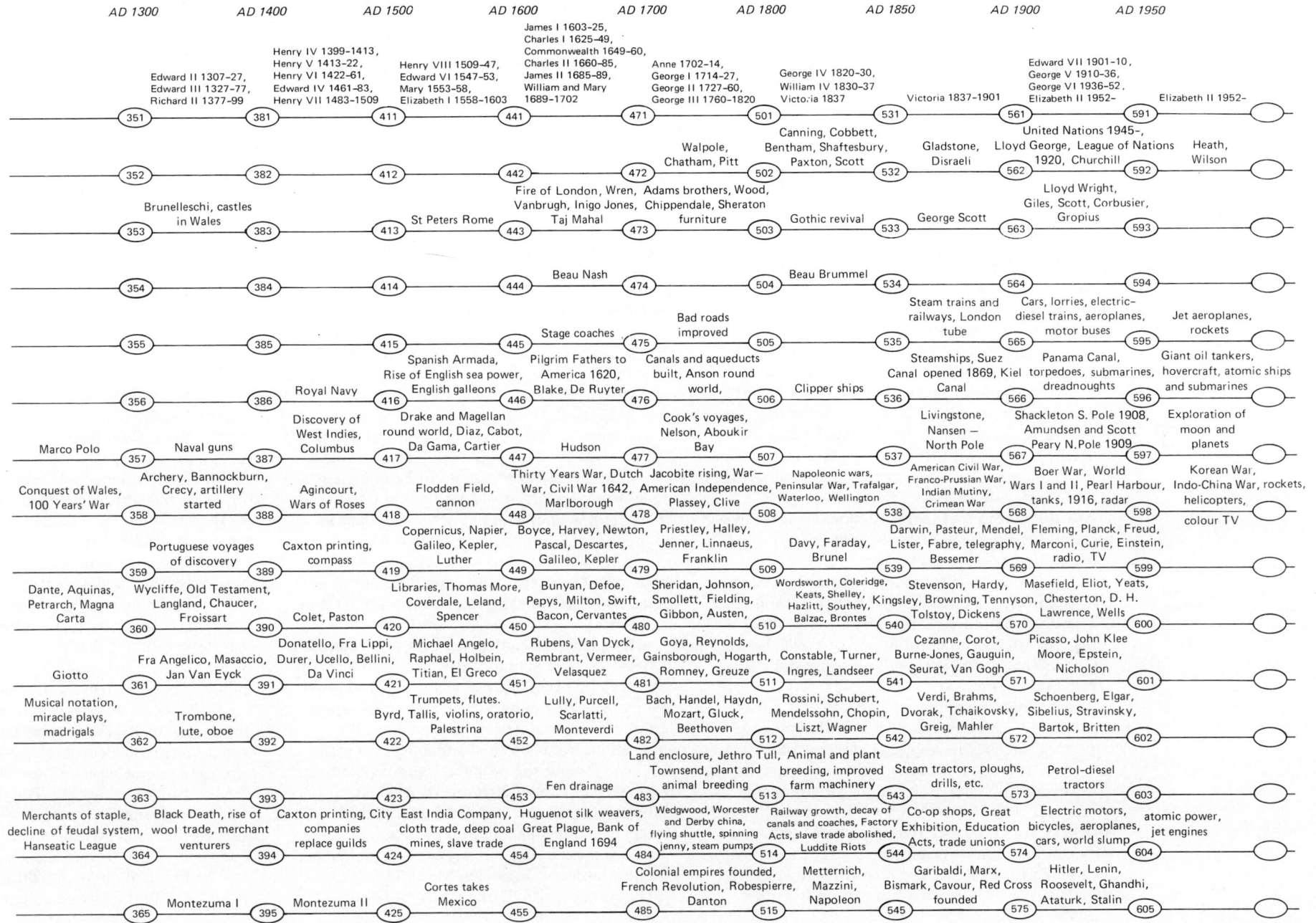

AD 1300　AD 1400　AD 1500　AD 1600　AD 1700　AD 1800　AD 1850　AD 1900　AD 1950

Edward II 1307–27,
Edward III 1327–77,
Richard II 1377–99

Henry IV 1399–1413,
Henry V 1413–22,
Henry VI 1422–61,
Edward IV 1461–83,
Henry VII 1483–1509

Henry VIII 1509–47,
Edward VI 1547–53,
Mary 1553–58,
Elizabeth I 1558–1603

James I 1603–25,
Charles I 1625–49,
Commonwealth 1649–60,
Charles II 1660–85,
James II 1685–89,
William and Mary
1689–1702

Anne 1702–14,
George I 1714–27,
George II 1727–60,
George III 1760–1820

George IV 1820–30,
William IV 1830–37
Victoria 1837

Victoria 1837–1901

Edward VII 1901–10,
George V 1910–36,
George VI 1936–52,
Elizabeth II 1952–

Elizabeth II 1952–

351　381　411　441　471　501　531　561　591

Walpole,
Chatham, Pitt

Canning, Cobbett,
Bentham, Shaftesbury,
Paxton, Scott

Gladstone,
Disraeli

United Nations 1945–,
Lloyd George, League of Nations
1920, Churchill

Heath,
Wilson

352　382　412　442　472　502　532　562　592

Brunelleschi, castles
in Wales

St Peters Rome

Fire of London, Wren,
Vanbrugh, Inigo Jones,
Taj Mahal

Adams brothers, Wood,
Chippendale, Sheraton
furniture

Gothic revival

George Scott

Lloyd Wright,
Giles, Scott, Corbusier,
Gropius

353　383　413　443　473　503　533　563　593

Beau Nash

Beau Brummel

354　384　414　444　474　504　534　564　594

Stage coaches

Bad roads
improved

Steam trains and
railways, London
tube

Cars, lorries, electric-
diesel trains, aeroplanes,
motor buses

Jet aeroplanes,
rockets

355　385　415　445　475　505　535　565　595

Royal Navy

Spanish Armada,
Rise of English sea power,
English galleons

Pilgrim Fathers to
America 1620,
Blake, De Ruyter

Canals and aqueducts
built, Anson round
world,

Clipper ships

Steamships, Suez
Canal opened 1869, Kiel
Canal

Panama Canal,
torpedoes, submarines,
dreadnoughts

Giant oil tankers,
hovercraft, atomic ships
and submarines

356　386　416　446　476　506　536　566　596

Marco Polo

Naval guns

Discovery of
West Indies,
Columbus

Drake and Magellan
round world, Diaz, Cabot,
Da Gama, Cartier

Hudson

Cook's voyages,
Nelson, Aboukir
Bay

Livingstone,
Nansen —
North Pole

Shackleton S. Pole 1908,
Amundsen and Scott
Peary N.Pole 1909

Exploration of
moon and
planets

357　387　417　447　477　507　537　567　597

Conquest of Wales,
100 Years' War

Archery, Bannockburn,
Crecy, artillery
started

Agincourt,
Wars of Roses

Flodden Field,
cannon

Thirty Years War, Dutch
War, Civil War 1642,
Marlborough

Jacobite rising, War—
American Independence,
Plassey, Clive

Napoleonic wars,
Peninsular War, Trafalgar,
Waterloo, Wellington

American Civil War,
Franco-Prussian War,
Indian Mutiny,
Crimean War

Boer War, World
Wars I and II, Pearl Harbour,
tanks, 1916, radar

Korean War,
Indo-China War, rockets,
helicopters,
colour TV

358　388　418　448　478　508　538　568　598

Portuguese voyages
of discovery

Caxton printing,
compass

Copernicus, Napier,
Galileo, Kepler,
Luther

Boyce, Harvey, Newton,
Pascal, Descartes,
Galileo, Kepler

Priestley, Halley,
Jenner, Linnaeus,
Franklin

Davy, Faraday,
Brunel

Darwin, Pasteur, Mendel,
Lister, Fabre, telegraphy,
Bessemer

Fleming, Planck, Freud,
Marconi, Curie, Einstein,
radio, TV

359　389　419　449　479　509　539　569　599

Dante, Aquinas,
Petrarch, Magna
Carta

Wycliffe, Old Testament,
Langland, Chaucer,
Froissart

Colet, Paston

Libraries, Thomas More,
Coverdale, Leland,
Spencer

Bunyan, Defoe,
Pepys, Milton, Swift,
Bacon, Cervantes

Sheridan, Johnson,
Smollett, Fielding,
Gibbon, Austen

Wordsworth, Coleridge,
Keats, Shelley,
Hazlitt, Southey,
Balzac, Brontes

Stevenson, Hardy,
Kingsley, Browning, Tennyson,
Tolstoy, Dickens

Masefield, Eliot, Yeats,
Chesterton, D. H.
Lawrence, Wells

360　390　420　450　480　510　540　570　600

Giotto

Fra Angelico, Masaccio,
Jan Van Eyck

Donatello, Fra Lippi,
Durer, Ucello, Bellini,
Da Vinci

Michael Angelo,
Raphael, Holbein,
Titian, El Greco

Rubens, Van Dyck,
Rembrant, Vermeer,
Velasquez

Goya, Reynolds,
Gainsborough, Hogarth,
Romney, Greuze

Constable, Turner,
Ingres, Landseer

Cezanne, Corot,
Burne-Jones, Gauguin,
Seurat, Van Gogh

Picasso, John Klee
Moore, Epstein,
Nicholson

361　391　421　451　481　511　541　571　601

Musical notation,
miracle plays,
madrigals

Trombone,
lute, oboe

Trumpets, flutes.
Byrd, Tallis, violins, oratorio,
Palestrina

Lully, Purcell,
Scarlatti,
Monteverdi

Bach, Handel, Haydn,
Mozart, Gluck,
Beethoven

Rossini, Schubert,
Mendelssohn, Chopin,
Liszt, Wagner

Verdi, Brahms,
Dvorak, Tchaikovsky,
Greig, Mahler

Schoenberg, Elgar,
Sibelius, Stravinsky,
Bartok, Britten

362　392　422　452　482　512　542　572　602

Fen drainage

Land enclosure, Jethro Tull,
Townsend, plant and
animal breeding

Animal and plant
breeding, improved
farm machinery

Steam tractors, ploughs,
drills, etc.

Petrol-diesel
tractors

363　393　423　453　483　513　543　573　603

Merchants of staple,
decline of feudal system,
Hanseatic League

Black Death, rise of
wool trade, merchant
venturers

Caxton printing, City
companies
replace guilds

East India Company,
cloth trade, deep coal
mines, slave trade

Huguenot silk weavers,
Great Plague, Bank of
England 1694

Wedgwood, Worcester
and Derby china,
flying shuttle, spinning
jenny, steam pumps

Railway growth, decay of
canals and coaches, Factory
Acts, slave trade abolished,
Luddite Riots

Co-op shops, Great
Exhibition, Education
Acts, trade unions

Electric motors,
bicycles, aeroplanes,
cars, world slump

atomic power,
jet engines

364　394　424　454　484　514　544　574　604

Montezuma I

Montezuma II

Cortes takes
Mexico

Colonial empires founded,
French Revolution, Robespierre,
Danton

Metternich,
Mazzini,
Napoleon

Garibaldi, Marx,
Bismark, Cavour, Red Cross
founded

Hitler, Lenin,
Roosevelt, Ghandhi,
Ataturk, Stalin

365　395　425　455　485　515　545　575　605

75

Fig. 44 (part 2).

7 · Network analysis in the development of the School Mathematics Project 7-13

An earlier chapter set out in some detail how an integrated scheme for primary mathematics was developed using a simplification of the network analysis technique as the basis. This chapter shows how a simplified network of the mathematics curriculum from age seven to thirteen was developed and then the manner in which teaching materials and other supplementary materials were designed, are now being tested in schools, and modified for publication from 1977 onwards. As this scheme is not yet complete all that can be discussed is what has been done to date. It is already becoming clear that this project will continue along the lines indicated here, though with some minor modifications for the older pupils to take account of their particular needs.

It will be seen that there are other areas where network analysis may be of value in this type of project. It is envisaged that there will be a large scale in-service training scheme for teachers. Since this will involve individuals and teams working in all parts of the United Kingdom and possibly in other countries, the co-ordination of this in-service work will require great care in planning and organisation.

In-service training for teachers of older pupils is concerned with a comparatively small number of specialist teachers, in comparatively few schools. In-service training for teachers of younger pupils involves all the teachers in a much larger number of schools, since there are very few specialist teachers. Any in-service training scheme for younger pupils must reach out to all the teachers in a much larger number of schools. Many of these teachers wish also to attend courses in the other subjects which they teach and so have to try to allocate the time they can spare for training, to all the subjects they teach. Many are married and have family commitments and so cannot attend extended courses held in big towns. Any attempt to reach the mass of teachers of younger pupils must consist of a large number of meetings held at local teachers' centres, or even at individual schools. The amount of work involved in planning this in-service training is immense, and it is when planning reaches this scale that network analysis proves invaluable.

I propose to deal with the planning and organisation involved in the School Mathematics Project 7-13 in some detail because it can be used as a model for other large-scale curriculum projects which are now in operation, or which might be planned for the future. Though network analysis is most valuable for large projects, smaller teachers' centre based projects might benefit from the technique. Even quite small projects can soon become complicated.

After a national conference held at Birmingham in May 1972, the School Mathematics Project was convinced that there was an urgent need to provide new materials for pupils below the age of eleven. There was then a gestation period while all the implications were absorbed and long-term plans were made to put the proposal into tangible form. The newly appointed director, Dr Alan Rogerson, visited a large number of schools, and talked to teachers, lecturers in colleges of education, advisers and others. As a result, twenty-five members of a new team met in September 1973.

The team consisted of teachers, or former teachers in primary, middle and secondary schools, mathematics advisers and lecturers in colleges of education. The team quickly set to work and reached broad agreement on the content of a mathematics curriculum for pupils from eight to thirteen years and an overall plan for producing teaching material.

Network design

At the first writers' meeting in September 1973 after a full discussion the content of the mathematics curriculum from age eight to thirteen was decided.

At the meeting a group allocated the items that made up the curriculum to a network framework so that each line of activities started at the level of simple discriminations, leading through multiple discriminations to concept formation. The activities included: three-dimensional shapes, weight, addition of number, plane shapes, subtraction, number, symmetry, time, multiplication of number, tessellation, volume, number bonds, rotation, recording data, topology, division of number, money, modular arithmetic, number bases.

A first draft was circulated to all the writing teams (figure 45). They suggested modifications which were then incorporated into the network which was used for the design of the teaching materials for the first years' testing. The two years' period for work was divided into six sections. It was envisaged that each division would correspond to about one term's work for the average pupil (figure 46).

The first column of activities was numbered from 1 to 50, the second column from 51 to 100

Network SMP 7–13

Fig. 45 — Curriculum network (node reference numbers shown in parentheses).

8 YEARS		9 YEARS			10 YEARS		11 YEARS → 13 YEARS			
Cayley tables (1)	Cayley tables (56)	Cayley tables (111)	Cayley tables (166)	Cayley tables (221)	Cayley tables (276)	(331)	(386)	(441)	(496)	(551)
'Posting' and rolling 3D shapes (2)	Fitting 3D shapes together (57)	Assembling 3D shapes to make models (112)	Cut clay shapes and examine (167)	Cut cartons and examine (222)	Make nets of cubes (277)	Make use of cuboids (332)	Assembling tetrahedra (387)	Assembling prisms (442)	Assembling icosahedra (497)	(552)
Ordering by weight using balance (3)	Ordering by weight using spring scales (58)	Graduating own scale of weight (113)	Graduating own scale of weight using standard units (168)	Graduating own scale of weight using standard units (223)	Graduating own scale of weight using standard units (278)	Compare units of weight and volume (333)	Compare combining weights (388)	Compare standard units and size (443)	(498)	(553)
Addition — number (4)	Addition — number (59)	Addition — number (114)	Addition problems (169)	Addition problems (224)	Addition problems (279)	(334)	(389)	(444)	(499)	Simple long division (554)
Sorting plane shapes by size (5)	Sorting plane shapes by corners (60)	Sorting plane shape and size (115)	Sorting similar shapes (170)	Sorting similar shapes of bilateral symmetry (225)	Sorting similar shapes of rational symmetrical properties (280)	Produce shapes on geoboard (335)	Produce shapes on geoboard (390)	(445)	Sorting prisms, etc. by size and shape (500)	Sorting prisms, etc. by size and shape (555)
Ordering by length (6)	Measuring using standard units (61)	Measuring using standard units (116)	Measuring with increasing discrimination (171)	Measuring with increasing discrimination (226)	Measuring with increasing discrimination (281)	Measuring with increasing discrimination (336)	Measuring with increasing discrimination (391)	(446)	(501)	(556)
(7)	(62)	(117)	(172)	(227)	(282)	(337)	(392)	(447)	Equivalent fractions (502)	(557)
(8)	(63)	(118)	(173)	(228)	(283)	(338)	(393)	(448)	Simple flow charts (503)	(558)
Subtraction as reverse of addition (9)	Subtraction as complementary addition (64)	Subtraction as complementary to addition (119)	Subtraction by decomposition (174)	Simple subtraction problems (229)	Simple subtraction problems (284)	Simple subtraction problems (339)	Rounding off (394)	Rounding off (449)	Simple inverse proportion (504)	(559)
Folding and cutting plane shapes along one axis (10)	Folding and cutting plane shapes along two axes (65)	Folding and cutting plane shapes along four axes (120)	Fold and cut plane shapes along four axes and compare magnetic compass points (175)	Fold and cut along axes plane shapes (230)	Fit plane shapes over each other (285)	Extend comparison of angles and sizes (340)	Tear off corners p/shapes fit tog. (395)	Explore angles of polygons (450)	Explore angles of polygons (505)	(560)
(11)	(66)	(121)	Compare clock face (176)	Compare clock face and earth rotation (231)	Compare seasons (286)	Compare seasons (341)	Compare moon and months (396)	Compare moon and months (451)	Explore past and future time (506)	(561)
Consolidate ability to tell time (12)	Compare various types of clock (67)	Use stop-watch to measure time and distance (122)	Use stop-watch to measure time and distance (177)	Use stop-watch to measure time and distance (232)	Use stop-watch to measure time and distance (287)	Compare speed and distance (342)	Compare speed and distance (397)	Compare speed and distance (452)	Compare bus-rail-air timetables (507)	(562)
Multiply two digits by one digit (13)	Multiply two digits by one digit (68)	Multiply three digits by one digit (123)	Multiply two digits by 10 (178)	Multiply three digits by 10 (233)	Multiply three digits by 10 (288)	Multiply three digits by 10 (343)	Multiply three digits by 10 (398)	Multiply three digits by 10 (453)	Multiply two digits by two digits (508)	(563)
Make patterns of regular and irregular plane shapes (14)	Simple tessellation (69)	Simple tessellation using squares and triangles (124)	Simple tessellation using rhombus (179)	Simple tessellation using quadrilaterals (234)	Simple tessellation using quadrilaterals (289)	Simple tessellation using quadrilaterals (344)	Tessellate H and L shapes (399)	Tessellate mixed regular p/shapes (454)	Tessellate curved shapes (509)	Tessellate irregular quadrilaterals and triangles (564)
Volume ordering using non-standard measures (15)	Volume-ordering using non-standard measures (70)	Volume-ordering using non-standard measures (125)	Explore standard measure in variety (180)	Explore standard measure in variety (235)	Explore standard measure in variety (290)	Explore standard measure with increasing discrimination (345)	Explore standard measure with increasing discrimination (400)	Explore standard measure with increasing discrimination (455)	Compare metric measure, weight and volume (510)	(565)
(16)	(71)	(126)	(181)	(236)	(291)	(346)	(401)	(456)	Use protractor to make clinometer (511)	(566)
Number bonds (17)	Commutative table squares (72)	Commutative table squares (127)	Table squares and associative properties (182)	Associative properties of table squares (237)	Introduce zero in addition (292)	Introduce zero in addition (347)	(402)	Explore addition and subtraction as reverse operations (457)	Explore multiplication and division as reverse operations (512)	Proportion (567)

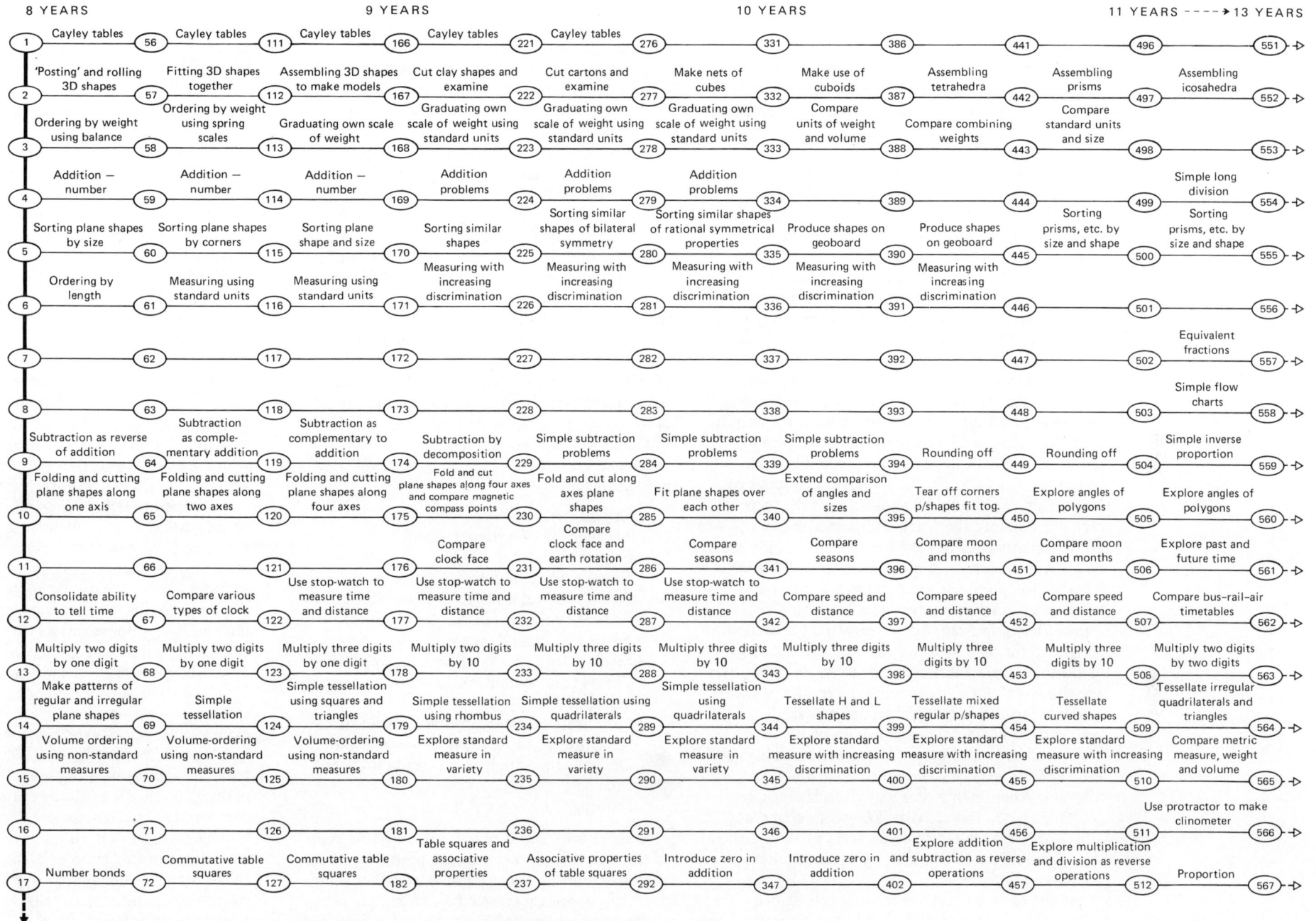

Fig. 45. *A section of a first draft of a curriculum of mathematics for infant-primary pupils, designed by a team for the School Mathematics Project 7–13. This design is based on the modification of network analysis by which each activity is given a unique reference number.*

and the third from 101 to 150. Activities 1, 51 and 101 are continuations of the same activity, becoming gradually more complex. Record cards were drafted for use in schools using this numbering, and were found to be effective and simple to use (figure 46).

This method of setting out the curriculum proved a great help when editing and allocating material. The numbering of each activity made for easy identification as it was sent in. Not only could it be sorted and placed in the correct order, but there was no possibility of any ambiguity. Each activity had a unique number. The number not only placed the activity in sequence along a line of activities of increasing complexity, but also positioned it in each column.

It was envisaged that a pupil would be placed by a class teacher in the column which best suited his needs, and that he could then select any activity in that column as a starting point. Having completed the activity he would then take his work to the teacher to be checked. Opinion varies as to whether pupils should be allowed to mark their own work before the teacher checks it. When the pupil presents his work the teacher can usually see whether it is correct. A question or two will elicit whether the pupil has grasped the point to be taught, and further teaching can be given if required. The pupil can then be directed to his next activity. This may be to repeat the same activity at another level of difficulty, or to go on to another activity. As the teacher records the results of the pupils efforts on his individual record card it is very easy to make a visual check on progress.

There were advantages for the writers in having each activity arranged in sequence on a network diagram. When each writer was given his task, he could easily see the activities he was to cover, and also how these related to other areas of the syllabus. Any dependence of one activity upon another could readily be seen.

At the early writers' meetings it was agreed that the teaching materials should not be graded, as this might tend to have an adverse effect upon the pupil. The material was not graded obviously in order of difficulty. Each card had a number at the foot but this was only for sorting purposes, and bore no relation to the degree of difficulty.

It was found that this added to the burden placed upon teachers, who almost without exception, wished the cards to be graded. They all felt that the 'labelling' of the pupils could be discounted. So the published versions of the teaching materials not only have the network number for identification but also some method of showing the mathematical progression or degree of difficulty of the material.

From the editing point of view the numbering system was a boon. There were hundreds of separate items of teaching material being processed, and had there been no simple method of following the progress of each separate sheet confusion would have resulted.

Each separate item of teaching material went through this process:

> Original writing
> Local group of writers critically examine and suggest alterations
> Rewrite where required
> Second critical examination by local group
> Critical editing by a team from another group
> Editing by central editors
> Sent to design team, after being typed out
> Design team submit rough drafts to central editorial team
> Modifications to rough draft, resubmitted to central editorial team
> Design team submit photo-ready copy for final edit
> Printing

As there were about one thousand items being fed through this process during a period of just over three months, it can be imagined that some means of checking was essential. Each item is numbered according to the network diagram, and for further identification each separate item of teaching material within an activity has a second number allocated. This is used for collation, and to enable the teachers to check that all the material is present. Thus cards for activity number 1 are: 1/1, 1/2, 1/3 and so on.

At an early stage the decision was taken to begin the project at age seven as this corresponds to the usual age of entry into junior schools in Britain. This meant a rapid reappraisal of the work to be done.

The first network was extended to take account of this new section and plans were made to have the necessary writing completed in time.

As the work proceeded it was realised that many teachers were bewildered by the sight of the network. At first glance it appears to be very confusing. Since one did not always have enough time to explain the layout of the network in sufficient detail to make it plain to all teachers who might be engaged in the project, it was decided to produce a simpler version.

It is open to debate how far it is necessary to detail what is in any curriculum. I think that for planning purposes, and those closely involved in a project, a detailed plan should be prepared which includes every activity. At some later stage this can be simplified so that the plan can be readily understood by those not so closely involved. For example I would suggest that editors of schemes of work, advisers and head teachers need a plan which contains all the relevant detail. The class teacher needs a simpler version which sets out the main points, but does not go into precise detail.

This in effect is what the SMP 7-13 project team decided. A meeting was called to revise the network so that it could be reduced in size and complexity. This was used for future work.

This revision in itself points to one of the advantages of this approach; the flexibility it gives those engaged in planning. Activities can be moved from place to place with ease. In addition the relationships between activities can be clearly seen

UNIT 1 UNIT 2 UNIT 3 UNIT 4

Strand (top — 3D shapes / Cayley tables): Cayley tables
- Making simple models from scrap — (51) Making simple models from bricks, boxes, tins, etc. — (101) Making simple models from clay. Examine faces, etc. — (151) 'Posting' and rolling 3D shapes — (201) Fitting 3D shapes together — (251) Assembling 3D shapes — making modes — (301) Cut clay shapes — explore — (351) Cut cartons — examine — (401) Make nets of cubes — (451) Make nets of cuboids — (501)

Strand (weight/volume):
- Weight — simple discrimination — (52) Personal scales — discrimination — (102) Increasing discrimination — (152) Ordering by weight using balance — (202) Ordering by weight using spring scales — (252) Graduating using own scale of weight — (302) Graduating using standard units — (352) Graduating using standard units — (402) Graduating using standard units — (452) Compare units of weight and volume — (502)

Strand (addition):
- Addition using structured apparatus — (53) Addition using structured apparatus — (103) Addition using structured apparatus — (153) Addition — number — (203) Addition — number — (253) Addition — number — (303) Addition problems — (353) Addition — number — (403) Addition — number — (453) — (503)

Strand (plane shapes sorting):
- Plane shapes sorting into sets of various properties — (54) Plane shapes sorting into sets of various properties — (104) Plane shapes sorting into sets of various properties — (154) Sorting plane shapes by size — (204) Sorting plane shapes by corners — (254) Sorting by shape and size — (304) Sorting similar shapes — (354) Sorting by bilateral symmetry — (404) Sorting by rational symmetrical properties — (454) Produce shapes on geoboard — (504)

Strand (length):
- Simple discrimination of length using non-standard units — (55) Simple discrimination using standard and non-standard units — (105) Increasing discrimination — (155) Ordering by length, width and height — (205) Measuring using standard units — (255) Measuring perimeter using standard units — (305) Measuring using increasing discrimination — (355) Measuring using increasing discrimination — (405) Measuring using increasing discrimination — (455) Measuring using increasing discrimination — (505)

- (56) (106) (156) (206) (256) (306) (356) (406) (456) (506)

- (7) (57) (107) (157) (207) (257) (307) (357) (407) (457) (507)

Strand (subtraction):
- Subtraction using structured apparatus — (58) Subtraction using structured apparatus — (108) Subtraction using structured apparatus — (158) Subtraction as reverse of addition — (208) Subtraction as complementary to addition — (258) Subtraction as complementary to addition — (308) Subtraction by decomposition — (358) Simple subtraction problems — (408) Simple subtraction problems — (458) Simple subtraction problems — (508)

Strand (sorting / folding plane shapes):
- Sorting plane shapes into sets — (59) Sorting plane shapes into sets and mapping — (109) Ordering sets — (159) Folding and cutting plane shapes along one axis — (209) Folding and cutting plane shapes along two axes — (259) Folding and cutting plane shapes along four axes — (309) Compare axes and magnetic compass points — (359) Fold and cut along axes — plane shapes — (409) Fit plane shapes over each other — (459) Extend comparison of angles and sizes — (509)

Strand (sets / Venn / clock):
- (60) Inclusive-exclusive sets Venn diagrams — (110) Intersecting sets — Venn diagrams — (160) (210) (260) (310) Clock face — (361) Compare clock face and earth rotation — (411) Compare clock face and seasons — (461) Compare clock face and seasons — (511)

- Sets by number — (61) (111) (161) (211) (261) (311) (361) (411) (461) (511)

Strand (time):
- Explore time measurement — (62) Explore days and months — (112) Explore hours and minutes — (162) Consolidate ability to tell time — (212) Compare various types of clock — (262) Use stop-watch to measure time and distance — (312) Use stop-watch to measure time and distance — (362) Use stop-watch to measure time and distance — (412) Use stop-watch to measure time and distance — (462) Compare speed and distance — (512)

Strand (multiplication):
- Multiplication using structured apparatus — (63) Multiplication using structured apparatus — (113) Multiplication using structured apparatus — (163) Multiply two digits by one digit — (213) Multiply two digits by one digit — (263) Multiply three digits by one digit — (313) Multiply two digits by 10 — (363) Multiply three digits by 10 — (413) Multiply three digits by 10 — (463) Multiply three digits by 10 — (513)

Strand (pattern / tessellation):
- Pattern making mosaics, etc. — (64) Drawing patterns, etc. — (114) Drawing squares and triangles. Play 'Battleships' — (164) Make patterns of regular and irregular plane shapes — (214) Simple tessellation — (264) Simple tessellation using squares and triangles — (314) Simple tessellation using rhombus — (364) Simple tessellation using quadrilaterals — (414) Simple tessellation using quadrilaterals — (464) Tessellate H and L shapes — (514)

Strand (volume):
- Volume — simple discriminations — (65) Volume — using standard and non-standard measures — (115) Increasing discrimination — (165) Volume ordering using standard measures — (215) Volume ordering using standard measures — (265) Volume ordering using standard measures — (315) Explore standard measures in variety — (365) Explore standard measures in variety — (415) Explore standard measures in variety — (465) Explore standard measures using increasing discrimination — (515)

Strand (counting / ordinal / commutative):
- Counting objects using structured apparatus, etc. — (66) Ordinal numbers using structured apparatus — (116) Ordinal numbers using structured apparatus — (166) Number bonds — (216) Commutative table squares — (266) Commutative table squares — (316) Commutative table squares and associative properties — (366) Commutative table squares and associative properties — (416) Introduce zero in addition — (466) Introduce zero in addition — (516)

(Left vertical strand nodes: 2, 3, 4, 5, 6, 7, 8, 9, 10, 11, 12, 13, 14, 15, 16)

(Right-hand exit nodes: 501, 502, 503, 504, 505, 506, 507, 508, 509, 510, 511, 512, 513, 514, 515, 516)

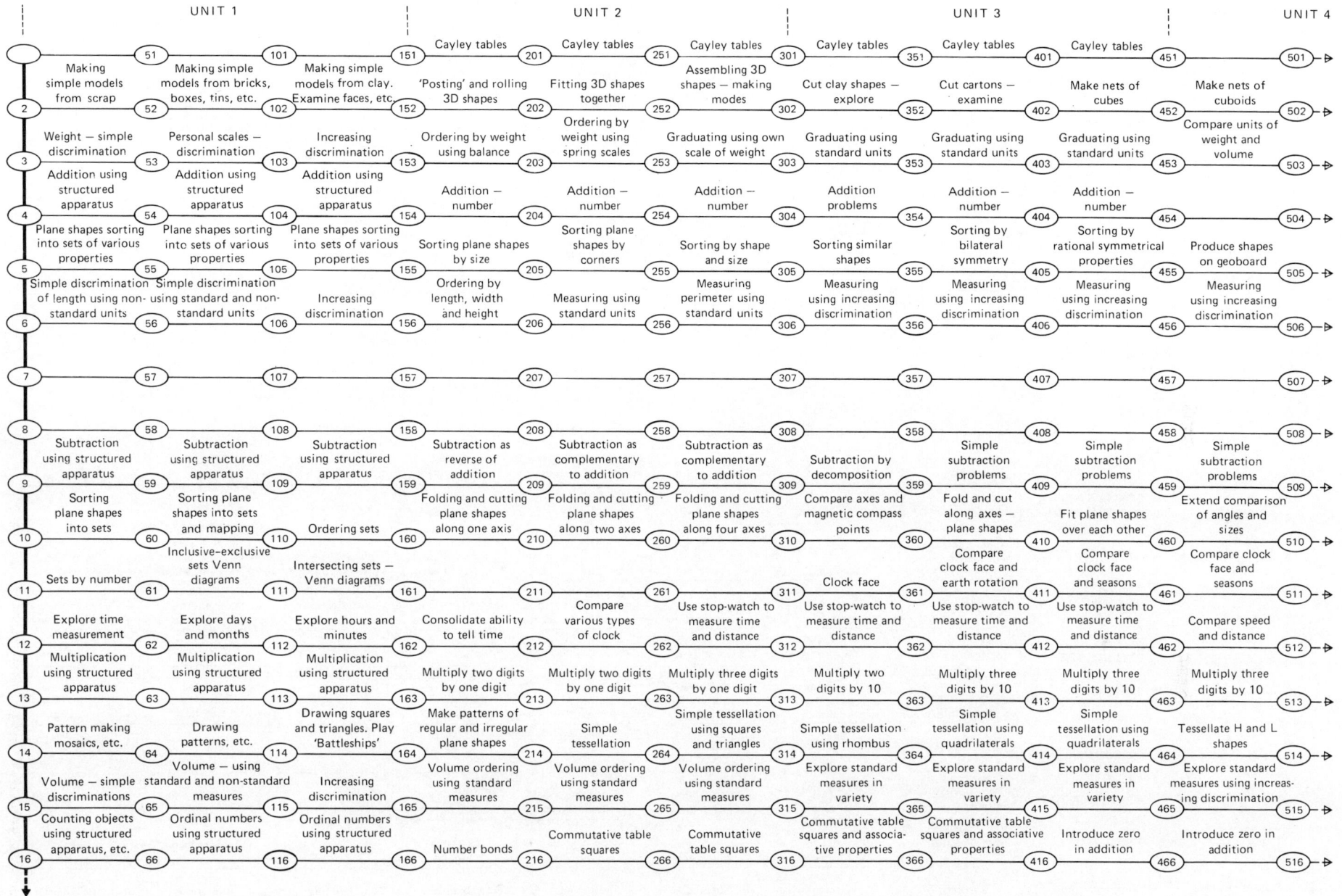

Fig. 46. *The second draft of the School Mathematics Project 7–13 mathematics curriculum. A section only is shown for ages 7–9 years.*

UNIT 1 UNIT 2 UNIT 3

Topic	Unit 1 (1)	(2)	(3)	Unit 2 (4)	(5)	(6)	Unit 3 (7)	(8)	(9)
SOLIDS	Making simple models from scrap	Simple models – identify edges	Simple clay models – identify edges	Posting-rolling solids	Fitting solids – Cubes and cuboids	Cubes and cuboids – making models	Cut clay solids – explore tetrahedra	Cut cartons, etc. explore. Explore cones cylinders *(Curve stitching using string to draw ellipses)*	Making nets of cubes. Explore building frameworks *(Use beam compass – radius and diameter; Explore compass patterns)*
WEIGHT	Simple discrimination – arbitrary units	Simple discrimination – ordering by weight	Increasing discrimination	Introducing single standard units	Introducing kilogram	Introducing simple computation	Measure metric units, use variety of balances, etc.	Measure metric units using personal scales	Measure metric units using spring scales
COMPUTATION ADDITION									
PLANE SHAPES	Sorting plane shapes into sets. Drawing-tracing	Making patterns using ruler	Ordering using ruler and set square	Sorting by size. Turning and tracing	Sorting by corners. Rotate about point in shape	Sorting shape and size continue rotation	Reflected symmetry folding: cut along one axis	Translational and bilateral symmetry. Fold and cut along two axes	Explorational symmetry. Rotate plane shape about point in shape. Cut along four axes
LENGTH	Simple discrimination – arbitrary units	Ordering by length	Ordering – increasing discrimination	Ordering by height, length, width, single standard unit	Introducing metres, centimetres	Measure perimeters – simple computation	Increasing discrimination	Increasing discrimination	Increasing discrimination
FRACTIONS	Folding, cutting into fractions ½ and ¼	Folding, cutting paper, beads, etc.	Further fractions	Further fractions	Further fractions	Explore equivalent fractions	Explore equivalent fractions	Explore equivalent fractions	Introduce percentage
COMPUTATION SUBTRACTION									
SETS	Mapping – one-to-one correspondence	Inclusive-exclusive sets – Venn diagrams	Intersecting sets – Venn diagrams	Ordering sets	Ordering sets	Ordering sets			
TIME	Explore time measurement before-after	Days, months	Hours, minutes	Consolidate ability to tell time	Compare various types of clock, 24-hr clock	Introducing stop-watch	Explore time-distance	Explore time-distance	Explore time-distance
COMPUTATION MULTIPLICATION									
AREA	Make patterns with regular and irregular plane shapes. Count squares in shapes	Simple tessellation potato prints. Draw shapes. Count squares	Simple tessellation using arbitrary squares, etc.	Superimpose grids on regular and irregular shapes	Area of plane shape using standard units	Area of plane shape using standard units	Use squares, triangles, etc. to find areas of plane shapes	Areas of regular and irregular plane shapes	Areas of regular and irregular plane shapes
VOLUME	Volume – simple discriminations	Various arbitrary units	Various increasing discrimination	Volume ordering introducing standard measure	Volume ordering introducing litre and ½ litre	Volume ordering introducing litre, 100 ml, etc.	Explore variety of metric measures	Explore variety of metric measures	Introduce four rules computation
NUMBER STRUCTURE	Number patterns, series. Number line-ladder. Domino type games	Ludo type games. Introduce number frame	Number line, frame, etc.	Introducing roman numbers	Introducing roman numbers	Introducing roman numbers	Associative properties	Associative properties	Associative properties
VERTICAL HORIZONTAL	Make simple level. plumb bob	Make simple level using marble. plumb bob	Continue with exploration of horizontal and vertical	Horizontal and vertical	Horizontal and vertical	Horizontal and vertical	Horizontal and vertical	Horizontal and vertical	Horizontal and vertical
COMPUTATION DIVISION									
PICTORIAL REPRESENTATION	Pictograms	Star charts	Explore tallies – block graphs	Collecting and recording data	Collecting and recording data	Tabulate data	Introduce line graphs	Introduce continuous numbers	Introduce continuous numbers
TOPOLOGY	Explore inside and outside plane shapes	Simple mazes	Clay mazes	Very simple networks	Colouring regions	Five colour problem	Explore regions, arcs and nodes	Euler's theorem	Produce topological transformations
MONEY	Shopping – one article	Shopping – two articles	Shopping – simple bills	Simple computation	Simple computation	Simple computation	Explore bills and budgets	Money problems	Introduce ready reckoners

(Node numbers across each line run 1–199: row stages are numbered 1, 21, 41, 61, 81, 101, 121, 141, 161, 181 for SOLIDS; 2, 22, 42, 62, 82, 102, 122, 142, 162, 182 for WEIGHT; and so on down to 19, 39, 59, 79, 99, 119, 139, 159, 179, 199.)

Fig. 47. *A further revision of the School Mathematics Project 7–13 mathematics. For pupils with mathematical ages 7-8-9 years.*

UNIT 4   UNIT 5   UNIT 6

Find radius of circles. Construct quadrilaterals.
Use ruler and compass to   Find diameter of   Use cogs to   Use cogs to show   Use cogs to show   Perimeter of
construct triangles (201)   circles (221)   show cycloids (241)   epicycloids (261)   epicycloids and pursuit curves (281)   (301)   circles (321)   (341)

Making nets of   Assembling   Assemble prisms.
cuboids. Explore rigid   tetrahedra. deforming squares.   Deform shapes.   Assemble   Assemble   Assemble   Filling spaces with   Fill cuboid with   Fill cuboid with
frameworks. Rotate solids   Rotate solids about an   Rotate solids about   icosahedra   dodecahedra   octahedra   solids — explore   different solids   different solids
about axis in shapes (202)   axis of symmetry (222)   axis not in shape (242)   (262)   (282)   (302)   (322)   (342)

Compare   Compare units of   Compare   Compare   Compare   Compare   Compare   Compare   Compare
units of weight and   weight and volume com-   standard units,   standard units,   standard units,   standard units,   standard units,   standard units,   standard units,
volume (203)   bining weights (223)   weight and size (243)   weight and size (263)   weight and size (283)   weight and size (303)   weight and size (323)   weight and size (343)   weight and size

(204)   (224)   (244)   (264)   (284)   (304)   (324)   (344)

Explore rotational   Explore and   Continue   Symmetrical   Angle   Congruent   Congruent
symmetry. Rotate plane   rotate plane shapes   Rotate plane shapes   Rotate plane shapes   exploration of regular   properties   properties   properties   properties
shapes about point in shape.   to make solids (225)   to make solids (245)   to make solids (265)   plane shapes (285)   of plane shapes (305)   (325)   (345)
Cut along four axes (205)

Increasing   Increasing   Increasing   Increasing   Increasing   Increasing   Increasing   Increasing   Increasing
discrimination (206)   discrimination (226)   discrimination (246)   discrimination (266)   discrimination (286)   discrimination (306)   discrimination (326)   discrimination (346)   discrimination

Introduce four   Explore directed
rules decimal   Increasing   and negative
computation (207)   (227)   discrimination (247)   numbers (267)   (287)   (307)   (327)   (347)

(208)   (228)   (248)   (268)   (288)   (308)   (328)   (348)

(209)   (229)   (249)   (269)   Explore   (289)   Explore   (309)   (329)   (349)
Explore time-   Compare   Compare   Compare   bus-rail-air   bus-rail-air   Explore   Explore   Explore
distance (210)   speed-distance (230)   speed-distance (250)   speed-distance (270)   timetables (290)   timetables (310)   acceleration (330)   deceleration (350)   deceleration

(211)   (231)   (251)   (271)   (291)   (311)   (331)   (351)
Introduce   Introduce regular   Introduce curved   Explore area   Compare similar   Compare similar   Pythagoras'
formulae areas   and mixed plane   shapes   solids   areas   areas   theorem
of plane shapes (212)   shapes (232)   (252)   (272)   (292)   (312)   (332)   (352)

Introduce and   Compare   Compare   Compare   Compare   Compare   Measure volume by   Measure volume   Measure volume by
compare units, weight   units, weights   units, weights   units, weights   units, weights   units, weights   displacement   by displacement   displacement
and volume (213)   and volume (233)   and volume (253)   and volume (273)   and volume (293)   and volume (313)   (333)   (353)

Make-use simple box
Associative   Make-use   Explore   Explore punch   punch card system. Explore   Ready   Explore: powers-   Explore   Explore
properties (214)   Napier's bones (234)   algorithms (254)   cards (274)   ready reckoners (294)   reckoners (314)   indices. Explore: punch card systems (334)   computers (354)   computers

Horizontal and   Horizontal and   Horizontal and
vertical (215)   vertical (235)   vertical (255)   (275)   (295)   (315)   (335)   (355)

(216)   (236)   (256)   (276)   (296)   (316)   (336)   (356)
Introduce   Mode. Introduce   Introduce   Explore scatter   Explore scatter   Explore scatter
pie graphs (217)   Mean (237)   averages (257)   median (277)   diagram (297)   diagram (317)   diagram (337)   (357)

Produce topological   Produce
transformations   topological
(218)   transformations (238)   (258)   (278)   (298)   (318)   (338)   (358)

Difference between   Difference between   Difference between   Explore rates   Explore rates   Explore rates   Explore
buying and   buying and selling   buying and selling price   and taxes   and taxes   interest rates
selling price (219)   price and profit (239)   and profit (259)   (279)   (299)   (319)   (339)   (359)

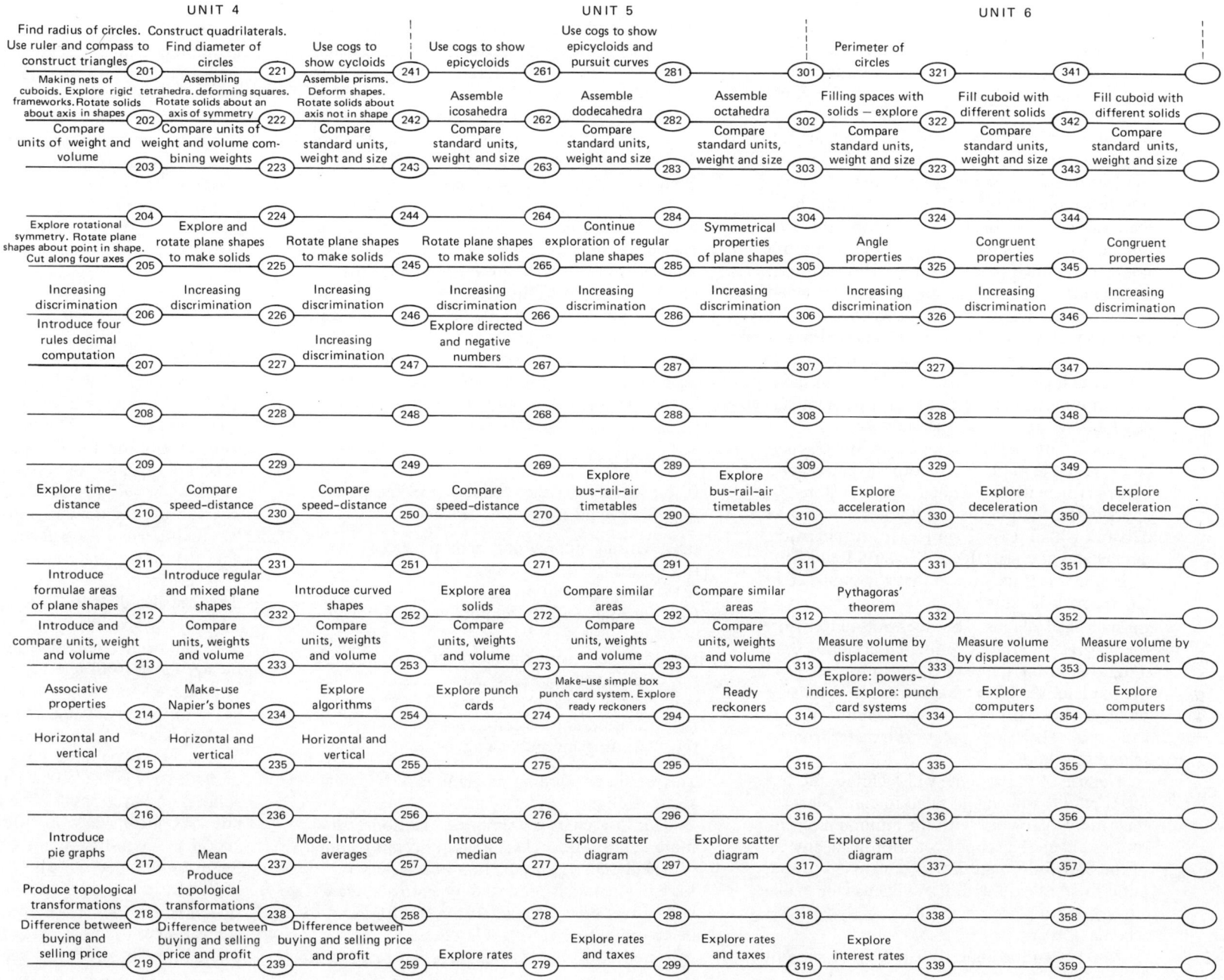

Fig. 47 (part 2). *This section for pupils aged 10–11–12 years.*

as they are all on one sheet. The dependencies of one activity upon another are also clear.

In the first network the work was sequenced to allow a variety of work for each of the years. Using the technique of the 'starter' card followed by a series of graded items of teaching material, it was hoped that there would be sufficient material to meet the needs of the average, the bright, and the slow pupils. Schools found that while the materials were reasonably satisfactory, there was not sufficient material for the bright pupils, nor was there enough practice for the slow learner.

When the syllabus was revised it was decided to arrange the syllabus so that activities in the first three columns were those more suitable for a pupil who was at mathematical age seven, regardless of his chronological age. The activities in the next three columns were for pupils of mathematical age eight and so on through the network.

This meant that in any one class the teacher could start the pupil at the point he had reached in his mathematical learning. This radical re-appraisal resulted in the production of a smaller network which may be simpler for teachers to understand and use (figure 47, parts 1 and 2).

In practice it may mean that the pupils in an unstreamed class may be quite widely spread through the syllabus. This does not matter since the progress of any individual pupil can be easily monitored using the record cards provided. When the pupil moves from class to class he can carry on his progress through the curriculum without any break. The same applies to transfer from school to school.

The use of this scheme will facilitate the transfer of pupils, and enable them to continue with their individual work with the minimum of disruption. Even if a school does not make any use of the SMP 7–13 materials a comparison of the pupil's record card with the syllabus will enable any school to place the pupil correctly within any scheme of work.

Apart from the benefits listed here I am quite

sure that all those who have been engaged in the design of the network diagram, and the revisions which have followed, will have gained new insights into the mathematics curriculum. The weighing of the correct sequence of teaching the various aspects of mathematics, and the examination of the dependences of one part upon another, makes the exercise of value for its own sake.

If at any time in the future it is necessary to make revisions in the scheme of work, then the logical, visual presentation of the whole syllabus on one sheet will facilitate this. Whether the revision is to add to the network fresh material which is not now there, or to remove items which prove to be unsuitable, or wrongly placed, the alteration will be quite simple. Any additions have provision built in, for there are spare numbers left in the network for just this eventuality.

Organisation

It was agreed that the production of teaching material should follow this process:

(a)   writing and revision in local groups
(b)   editing
(c)   designing and printing
(d)   testing in a limited number of schools for one year
(e)   revision, using the feedback from schools
(f)   re-printing as necessary
(g)   testing for another year
(h)   revision as necessary
(i)   printing and publishing

This would be a rolling programme in two-year sections (figure 48).

This flow chart is a simplified version of the more detailed flow chart used in overall planning.

The design and production of a variety of draft teaching materials produced by a writing team widely scattered throughout the country, had to be co-ordinated. Testing schools, also widely scattered, had to have draft materials in good time

for the two-year testing programme. The design team producing draft materials and printers were in close touch with the editors and their work had to be planned and woven into the whole scheme.

In order to keep the whole project on course, and to act as a means of giving information to all those engaged, a more detailed flow chart was designed (see figure 49).

From this flow chart it can be seen that there are a series of parallel activities being carried out by different teams. These parallel activities have to be kept in step, and regular meetings were held so that the teams could discuss all problems.

Teaching materials

An early decision was taken that the teaching materials should be in the form of work cards and topic books. It was thought that material in book form did not meet the needs of pupils in the variety of school organisations which can now be found in this country. Most of the schools try in some way to cater for the individual needs of pupils. For their needs a book is not flexible enough as a unit of teaching material.

Apart from the variety of school organisations there are also differences in age of transfer from school to school in different parts of Britain.

It had been agreed that the scheme should cover the age range from seven to thirteen and it should be planned so far as possible that pupils would not suffer in any change from school to school.

Any scheme based on books tends to be used on a one book per class basis. A variety of grades of material was needed because of the variety in abilities amongst pupils. The shorter work card printed on one side would meet the needs of the majority of children at age seven but more work for those who need it can be printed on the backs of the cards. Longer 'topic books' have been included where the nature of the work being done seems to merit a more extended treatment, or where more illustration is needed.

Since the syllabus is designed as a continuous stream, the materials follow the same developmental pattern. The numbering of the activities on the network enables the teaching materials to be slotted into the overall scheme with the maximum efficiency.

The size of the cards was fixed as A5 (21 x 15 cm) as this is convenient to handle by children, and will fit inside their exercise books. It is easy to store on a classroom shelf. It contains enough material for a seven-year-old pupil of average ability and with the right amount of illustration. Once the size of the teaching materials had been decided it was easy to design a method of storing which was simple and effective. Once again the numbering system of the network of the syllabus makes the storage and retrieval of the materials simple and easy to use, even for children of seven. Boxes can be made compact and easy to store and move about if necessary.

As the cards will be in constant use they have to be made of durable materials. The published version is printed on stiff card coated to keep the surface clean and prolong the life of the materials. This is not only common sense and an obvious economy in the long run, but also gives the pupils a pride in using good quality materials.

There were some differences of opinion about the typeface to be used. It is quite clear that for younger children it is necessary to use a larger typeface, but the question was which sort of letter shape do children prefer, or would they find it easier to read handwritten material? In many schools now pupils are used to handwritten materials prepared by their teachers.

The draft materials for use in testing schools were designed to find out what would be preferred. The final versions use a mixture of typography and handwriting as a result of these tests.

Evaluation

In the planning stage it was thought wise to ap-

point someone to engage in continuous evaluation of the whole project. It was decided to limit the number of schools involved in testing the draft teaching materials to thirty. It was thought that this many schools giving careful evaluations would produce better results than large numbers giving a vast amount of possibly less well considered opinions. In addition it was possible for the testing schools to be visited at regular intervals to see what was happening, and for informal talks with teachers and head teachers. Many are prepared to

voice valuable comments, which they might not be so ready to commit to paper. Visitors also may see snags which may not be mentioned. The traffic is, of course, two-way as editors may sometimes be able to suggest how difficulties may be overcome, or prevented from arising.

Each teacher is asked to make a daily note of any items which could be improved. At the end of each term a full report based on a questionnaire is sent in by each teacher at all the testing schools. Once each year the pupils' record cards from each

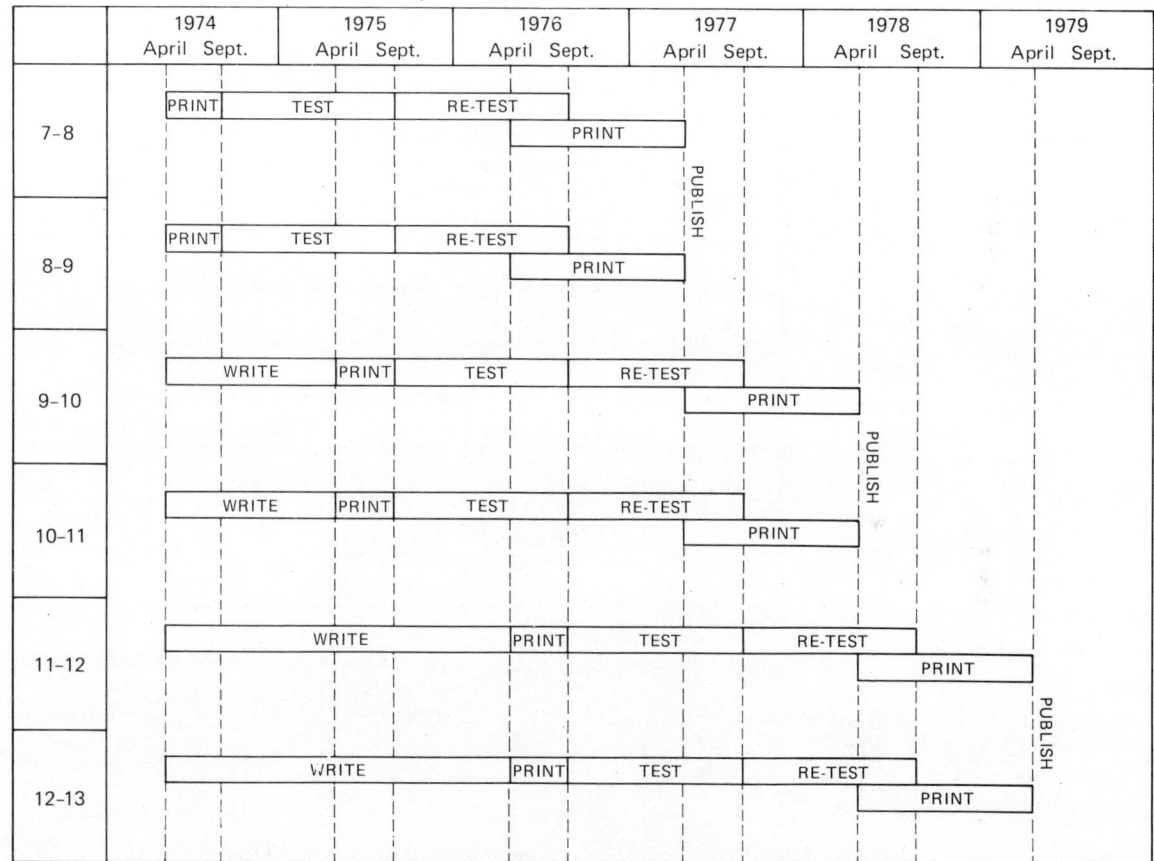

Fig. 48. *Simplified flow chart of the planning of the School Mathematics Project 7–13 teaching materials.*

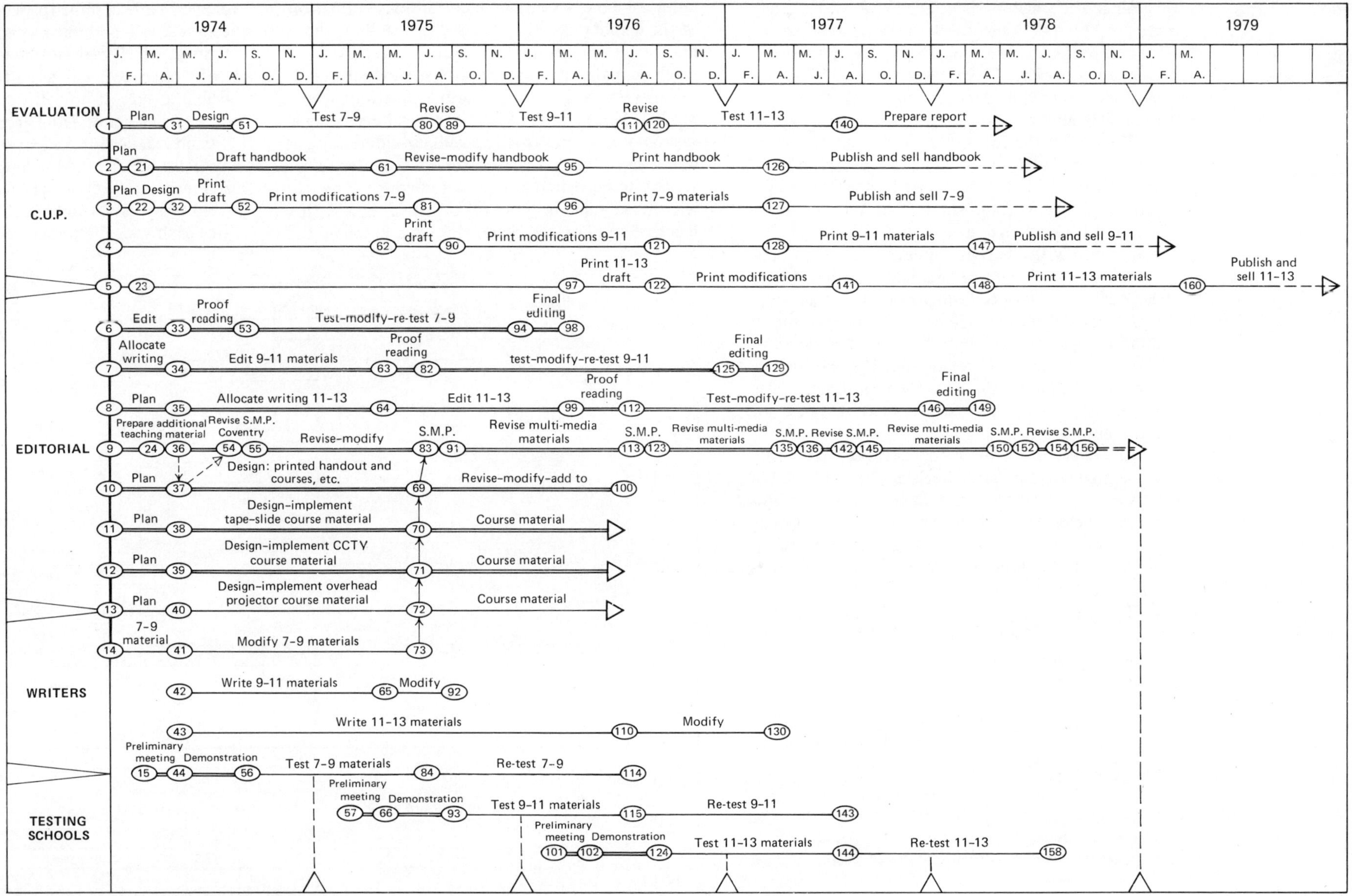

| | 1974 | 1975 | 1976 | 1977 | 1978 | 1979 |

| | J. F. | M. A. | M. J. | J. A. | S. O. | N. D. | J. F. | M. A. | M. J. | J. A. | S. O. | N. D. | J. F. | M. A. | M. J. | J. A. | S. O. | N. D. | J. F. | M. A. | M. J. | J. A. | S. O. | N. D. | J. F. | M. A. | M. J. | J. A. | S. O. | N. D. | J. F. | M. A. |

EVALUATION
1 — Plan — 31 — Design — 51 — Test 7–9 — Revise 80 89 — Test 9–11 — Revise 111 120 — Test 11–13 — 140 — Prepare report

C.U.P.
2 — Plan — 21 — Draft handbook — 61 — Revise–modify handbook — 95 — Print handbook — 126 — Publish and sell handbook
3 — Plan Design — 22 32 — Print draft — 52 — Print modifications 7–9 — 81 — 96 — Print 7–9 materials — 127 — Publish and sell 7–9
4 — 62 — Print draft — 90 — Print modifications 9–11 — 121 — 128 — Print 9–11 materials — 147 — Publish and sell 9–11
5 — 23 — 97 — Print 11–13 draft — 122 — Print modifications — 141 — 148 — Print 11–13 materials — 160 — Publish and sell 11–13

EDITORIAL
6 — Edit — 33 — Proof reading — 53 — Test–modify–re-test 7–9 — 94 — Final editing — 98
7 — Allocate writing — 34 — Edit 9–11 materials — 63 — Proof reading — 82 — test–modify–re-test 9–11 — 125 — Final editing — 129
8 — Plan — 35 — Allocate writing 11–13 — 64 — Edit 11–13 — 99 — Proof reading — 112 — Test–modify–re-test 11–13 — 146 — Final editing — 149
9 — Prepare additional teaching material — 24 36 — Revise S.M.P. Coventry — 54 55 — Revise–modify — 83 91 — S.M.P. — Revise multi-media materials — 113 123 — S.M.P. Revise multi-media materials — 135 136 142 145 — S.M.P. Revise S.M.P. Revise multi-media materials — 150 152 154 156 — S.M.P. Revise S.M.P.
10 — Plan — 37 — Design: printed handout and courses, etc. — 69 — Revise–modify–add to — 100
11 — Plan — 38 — Design–implement tape–slide course material — 70 — Course material
12 — Plan — 39 — Design–implement CCTV course material — 71 — Course material
13 — Plan — 40 — Design–implement overhead projector course material — 72 — Course material
14 — 7–9 material — 41 — Modify 7–9 materials — 73

WRITERS
42 — Write 9–11 materials — 65 — Modify — 92
43 — Write 11–13 materials — 110 — Modify — 130

TESTING SCHOOLS
15 44 — Preliminary meeting Demonstration — 56 — Test 7–9 materials — 84 — Re-test 7–9 — 114
57 66 — Preliminary meeting Demonstration — 93 — Test 9–11 materials — 115 — Re-test 9–11 — 143
101 102 — Preliminary meeting Demonstration — 124 — Test 11–13 materials — 144 — Re-test 11–13 — 158

Fig. 49. *Detailed flow chart of the planning involved in the production of the School Mathematics Project 7–13 materials.*

school are sent in for analysis. These show what each pupil has done, and also how well he has done.

Although there is a considerable mass of data to be processed by the evaluator, it is proving to be manageable. The results of the evaluation are fed back to the writers and used to improve not only the whole scheme but also the individual items of teaching material.

The publishers

Cambridge University Press are publishing the scheme as it completes the testing process. Their editor responsible has been at every writers' meeting and other planning meetings. All the work done has been in full co-operation with him, and with his helpful advice. He has made a large number of valuable contributions to discussions especially when questions of actual production processes have been involved. Representatives from CUP are also present at every meeting involving the SMP and the design team working on the draft material.

Editors of SMP 7–13

The SMP editors have a central place in the scheme. Apart from editing the teaching materials as they come from the writers, they are involved in every other aspect of the project including meeting the need for back-up materials for in-service training to be used as soon as the first two years of material are published.

There was considerable discussion as to how far we could proceed with the provision of teaching materials especially designed to meet the needs of the slow learners, and this is still under review as this book is being completed. There is no doubt that there is a real need for such provision but the difficulties are considerable. There might be a need for a large amount of material in very small graded steps, and the cost for a small proportion of the

Writers

I have already mentioned the variety of backgrounds of the members of the writing team. This might have caused dissent, but they quickly knitted into a hard-working team. There are several groups of writers who meet regularly in different parts of the country. The editor sends them a section of the network after prior discussion. The group then meets to decide which parts shall be tackled by each individual writer. When they have produced a sample this material is submitted to the critical scrutiny of the group. The writer then makes such alterations as appear to be needed. This material is re-submitted to group scrutiny before sending off to the central editor. From time to time the central editor is able to visit a group and discuss problems and make constructive suggestions so that there is an overall consistency in the team's work. At the end of the period allotted to writing all the materials are gathered at a central writers' meeting. There different writers critically examine the materials produced by another group. Further amendments are made if necessary. Finally the material is handed over for the central editor to check before handing over to the design team.

Testing

Before schools decide to test the materials, quite naturally they need to be convinced of their value. This means that the central editorial team pays a visit or several visits either to a teachers' centre, or a central school, to which staff from nearby schools are invited. After a talk explaining the scheme, and a display of samples of the teaching materials, questions are invited. Sometimes the talk and display is supplemented by a practical demonstration of how the materials can be introduced into the classroom, using a class of pupils. The testing has always been with the full co-operation

school population could be as great as for the rest of the school.

of the head and staff, after a series of explanatory meetings. The local mathematics advisers also have been consulted at every stage and have given valuable and active support.

Teaching materials – deployment

In the early stages it was thought that it would be best to meet the individual needs of the pupils by having a branching programme of the type set out in the diagram in figure 50.

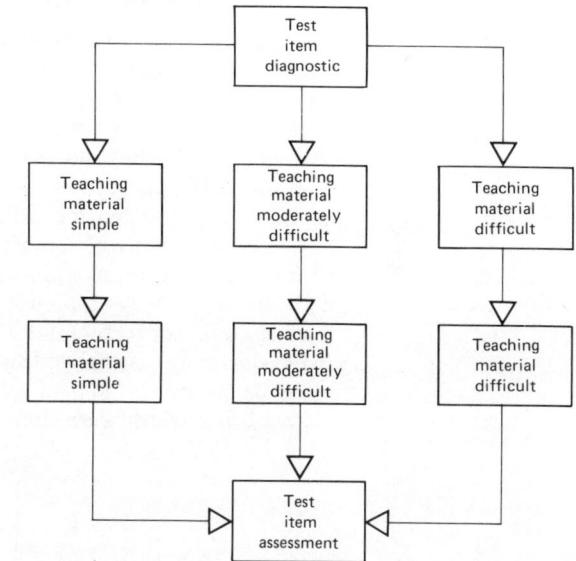

Fig. 50. *Suggestion for teaching material design to be used in School Mathematics Project 7–13. This was rejected because of the complexity of designing such material and use in the classroom.*

When this idea was examined more closely some difficulties came to light.

1   How many branches ought one to design in order to meet the needs of all the pupils in an unstreamed class?

2   If a pupil failed to pass one or more of the test items, should there be a further sequence, or a remedial sequence, and if he failed again, another?

3   How could this be administered by busy teachers?

4   Was there a way of keeping track of the progress of the individual pupil?

5   If the attention span of pupils is between ten and fifteen minutes on average, how long ought they to be engaged on the branching programme?

On the whole it was thought that although such branching programmes might possibly be effective, the design and testing of such a complicated series of learning situations for the wide range of pupils envisaged was far beyond our capacity.

A much simpler form of deployment was used in which there is a 'starter' card. This was intended to be a simple card which most pupils could tackle with confidence and would increase motivation with success. If they could not do the starter, the teacher could try to find out why and give help. If the starter was completed successfully, the pupil could be guided to the next activity. After much testing starter cards were later discarded in favour

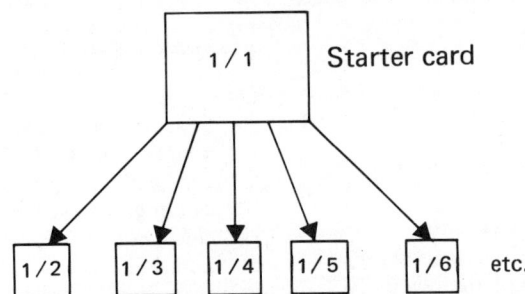

Fig. 51. *Arrangement of draft teaching materials of School Mathematics Project 7–13 which were tested out and rejected. The 'starter' card had a picture of an animal to distinguish it.*

of a series of items of increasing difficulty in mathematical progression (figure 51).

Although it was thought that a combination of shorter work cards with more extended topic books would best meet the needs of pupils most of the time, writers were sure that there was a good case to be made for group and class work. There is a definite need for bringing together pupils in larger groups to discuss common errors or difficulties, and for engaging in work jointly with others. To this end a number of special cards were included at intervals which set out some ways in which extension work should be done by teachers and pupils together, either in class or smaller groups. The value of this sort of inclusion was thought to be not only to avoid the repetition involved in teaching common errors but also in the social value of working with others as a group.

As all the teaching materials are numbered according to the network of the syllabus there is no problem in deploying such group or classwork cards with the rest of the materials.

Record cards

Once a number of pupils start to work as individuals an effective method of record-keeping becomes essential.

Since all pupils start at different points in the syllabus, and proceed at different paces through it there is a wide range of attainment and this varies as the pace of individual pupils varies with their health, emotional stability and perseverence. In schools which have larger groups of children working in areas, the problem of monitoring the progress of the individual child becomes almost insuperable.

What is needed is a simple but effective method of record-keeping. It must be simple so as to take up the minimum of teaching time. It should show what the child has done, and at what level he has tackled the work, together with his success or failure. It should for preference be on one side of

one sheet. It should show at a glance the point in the overall scheme he has reached.

Schools vary very much in their use of records. It could be that some have not examined very closely the reasons why records are needed. In some schools there are no records kept at all beyond class lists; in one I was told that at the end of each day each teacher sat down and wrote an account of what each child had done that day!

When classes were streamed, the teacher taught through the syllabus and at the end of the term could say that all the pupils had covered so much of the course. Once pupils start working as individuals, then it is necessary to record what each has done, and at what level of difficulty.

Once pupils start on an individual course then record cards are needed for the reasons which have been set out in earlier chapters.

The draft record cards designed to be used with the SMP scheme were based upon the numbering system used in the network drawing of the syllabus. Each item on the syllabus is numbered in sequence, and each item has a unique number allotted to it. Figure 52a is a section from the final version of the record card, now published.

SMP 7-13 Unit 2   **Name** _____

School Number _____

151/1	159/9	166/6	176/2	185/2	206/7	218/3	228/6	235/1
152/1	160/1	7	3	3	8	4	7	2
2	2	167/1	4	4	209/1	219/1	8	3
153/1	3	168/1	178/1	5	2	2	9	4
2	4	2	2	186/1	3	3	229/1	5
3	161/1	3	3	2	4	220/1	2	236/1
4	2	4	4	187/1	5	221/1	3	2
5	162/1	169/1	5	2	6	2	4	3
6	2	2	180/1	3	7	3	230/1	4

Fig. 52a. *A section from an SMP 7–13 record card tested out.*

SMP 7-13 Unit 1   Record Sheet   Name _____   Class _____

Addition	1— 1 2 3 4 5 6 7 8 9 10 11 12 13 14 15 16 17 18 19 20 21 22 23 24 25 26 27 28 29 30 31 32 33 34 35 36 37 38 39 40 41 42 43 44 45 46 47 48 49 50	21— 1 2 3 4 5 6 7 8	41— 1 2 3 4 5 6 7 8
Subtraction	2— 1 2 3 4 5 6 7 8 9	22— 1 2 3 4 5 6 7 8 9 10 11 12	42— 1 2 3 4 5 6 7 8 9 10
Multiplication	3— 1 2 3	23— 1 2 3	43— 1 2 3 4 5 6 7 8 9 10 11 12 13 14 15 16 17 18 19 20 21 22 23 24 25 26 27 28 29 30 31 32 33 34 35 36 37
Division	4— 1 2 3 4	24— 1 2 3	44— 1 2 3 4
Fractions	5— 1 2 3 4	25— 1 2 3 4	45— 1 2 3 4 5 6
Money	6— 1 2 3 4 5 6	26— 1 2 3 4 5 6 7	46— 1 2 3 4 5 6 7 8 9 10
Length	7— 1 2 3 4 5	27— 1 2 3 4 5 6 7	47— 1 2 3 4 5 6 7 8
Area	8— 1 2 3 4 5	28— 1 2 3 4 5	48— 1 2 3 4 5 6 7 8 9
Volume and capacity	9— 1 2 3 4 5 6 7 8 9 10	29— 1 2 3 4 5 6 7 8	49— 1 2 3 4 5
Weight	10— 1 2 3 4 5	30— 1 2 3 4 5	50— 1 2 3 4 5 6 7
Time	11— 1 2 3 4 5 6 7	31— 1 2 3 4 5 6 7 8 9 10 11 12	51— 1 2 3 4 5 6
Graphs	12— 1	32— 1	52— 1
Shape	13— 1 2 3 4 5 6 7 8 9 10	33— 1 2 3	53— 1 2 3 4 5 6
Angles	14— 1 2 3 4 5 6	34— 1 2 3 4	54— 1 2 3 4 5 6 7
Assessment tests			

Fig. 52*b. Later published version of record card.*

It is not only necessary to know what a pupil has done but also how many times he has carried out an activity. Some pupils quickly grasp new ideas and can rapidly pass from one activity to another. Others need a lot of repetition and reinforcement. This should have a place on the record card; on the SMP record cards there is a small square opposite each number for such entries.

The success or failure of the pupil can be shown easily by making different marks. For example, the correct completion of an activity can be shown as an 'O' and the incorrect as an 'X'.

Another piece of information which can be shown on the record card is the level of difficulty of the activity. The feedback from testing schools was that teachers wanted the materials to be clearly graded in order of difficulty, or mathematical progression, and this has been done with the published materials.

The numbering system enables a teacher to follow the progress of each pupil easily and check it at a glance. It ensures that the pupil covers the whole course and does not just pick out the bits he fancies at the expense of others which may be less appealing. It shows how many times he completes each activity and at what level of difficulty.

The records are kept on one side of the card and can be arranged in alphabetical order, or in any other order that suits the teacher. Each entry takes only a few seconds and so does not occupy valuable time. The pupil may be told to find his own record card and take it to his teacher along with his completed work for checking, saving more time.

In later published teaching materials a simplified version of a record card has been included in each self-contained unit (figure 52*b*).

The future

All that has been written about the SMP 7–13 Project applies to the early stages of planning. It will be realised that there have already been a

number of changes made. This is due to a deliberate policy of introducing some aspects of work to be tested, so that the reaction of teachers and pupils may be obtained before publication. Thus after the first years' testing in schools all comment from head and class teachers was carefully evaluated, and alterations made to the first two years' material which would be tested for another year. The same alterations were also made in the third years' material which is due to be tested for two years. This is an on-going process, of testing out material, modifying and re-testing, which has been embodied in the project from the start.

We are fortunate to have an organisation like the School Mathematics Project which has been able to develop this sort of approach, and has the will, the foresight, and ability to gather teams to carry out the work involved.

Postscript by Dr Alan Rogerson, Director of the SMP 7–13 project

This chapter captures in vivid detail the early discussions and planning involved in the SMP 7–13 project. Even though many of our decisions have been modified and improved following the testing in schools, it is useful for those interested in the *process* of curriculum development to see the preliminary planning, and the 'scaffolding' as it were, before the final building is finished. In this respect the *principle* of network planning which was very much Bill Vaughan's contribution to SMP 7–13 proved to be a valuable (even essential) mechanism. It helped not only with the overall planning but a much simplified version of the curriculum network emerged as a major asset to teachers and children using the scheme.

Work on SMP 7–13 will continue right through to 1980. For up to date details please write to the School Mathematics Project, Westfield College, Kidderpore Avenue, London NW3 7ST or Cambridge University Press.

8 · School organisation

In this chapter I explain how the operational research technique of network analysis may be used as an aid to planning, organisation and scheduling in education. I give examples of a variety of ways and situations in which the technique or modifications of it can be used. It could be argued that it would have value in the following: schools, colleges of education, colleges of further education, schools or faculties at universities and apprentice training. Educational establishments have tended to increase in size with courses increasing in scope and variety. Increases in size and complexity call for more powerful methods of planning than the somewhat informal ones that sufficed in smaller and less complex establishments and situations. Large-scale events in the educational year can be planned by means of a network and updated from year to year to make for greater efficiency, especially when staff turnover is rapid. New methods of teaching call for more effective planning and organisation if the pupils are to be guided and monitored effectively.

There can be no doubt that new methods of controlling large educational establishments in an effective and economical manner are vitally important. Any method which promises to give greater control, or save manpower, time or other resources must be given a trial, especially when it has proved to be effective in other fields.

The school year

In any form of planning the amount of detail required will vary and depend upon the planner's level of responsibility. A headmaster will need to have a very clear plan of the overall work of the school but need not necessarily be involved in the detailed planning of each series of activities; indeed it would not be possible for the head in a large school to be so involved. A subject specialist will need an overall plan so as to be able to relate what his department is doing with the work of the rest of the school.

A basic plan for the headmaster of a large school might simply be a flow chart as shown in figure 53.

If the flow chart in figure 53 is broken down into more detail it might resemble the network analysis in figure 54.

The school year starts in September, but activities which start in September have to be planned in June and July. The network in figure 54 starts with the calendar year in January. Not all the activities will apply to all schools, and there will be activities in some schools, which are not included in the network. However, the example here can be regarded as a general model for individual schools to adapt to their particular circumstances.

The line of activities involving stock checking,

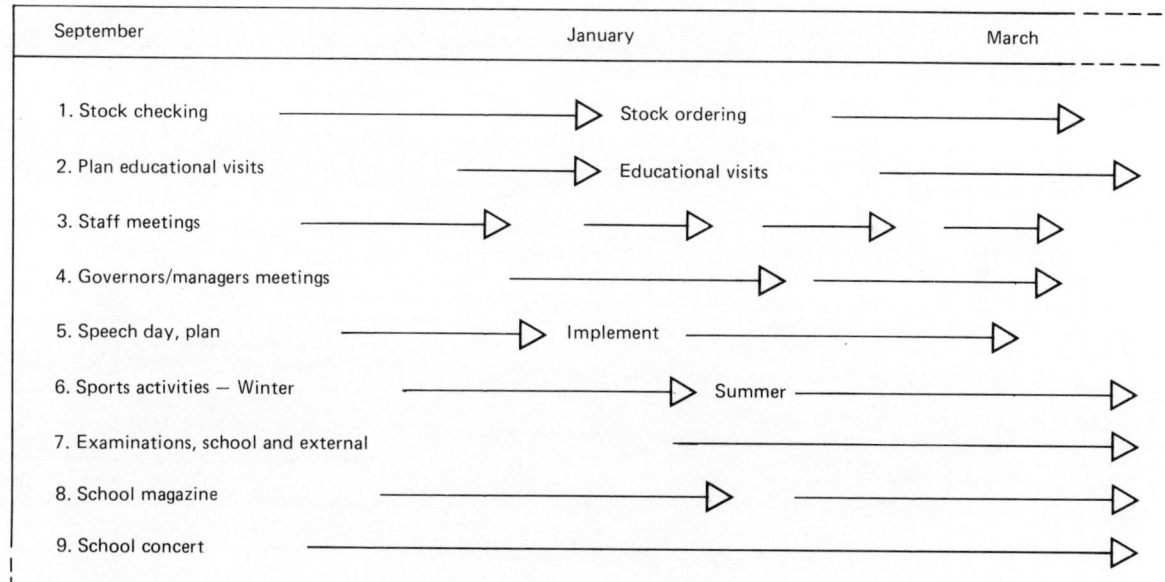

Fig. 53. *A simple flow chart of some of the annual events in a school.*

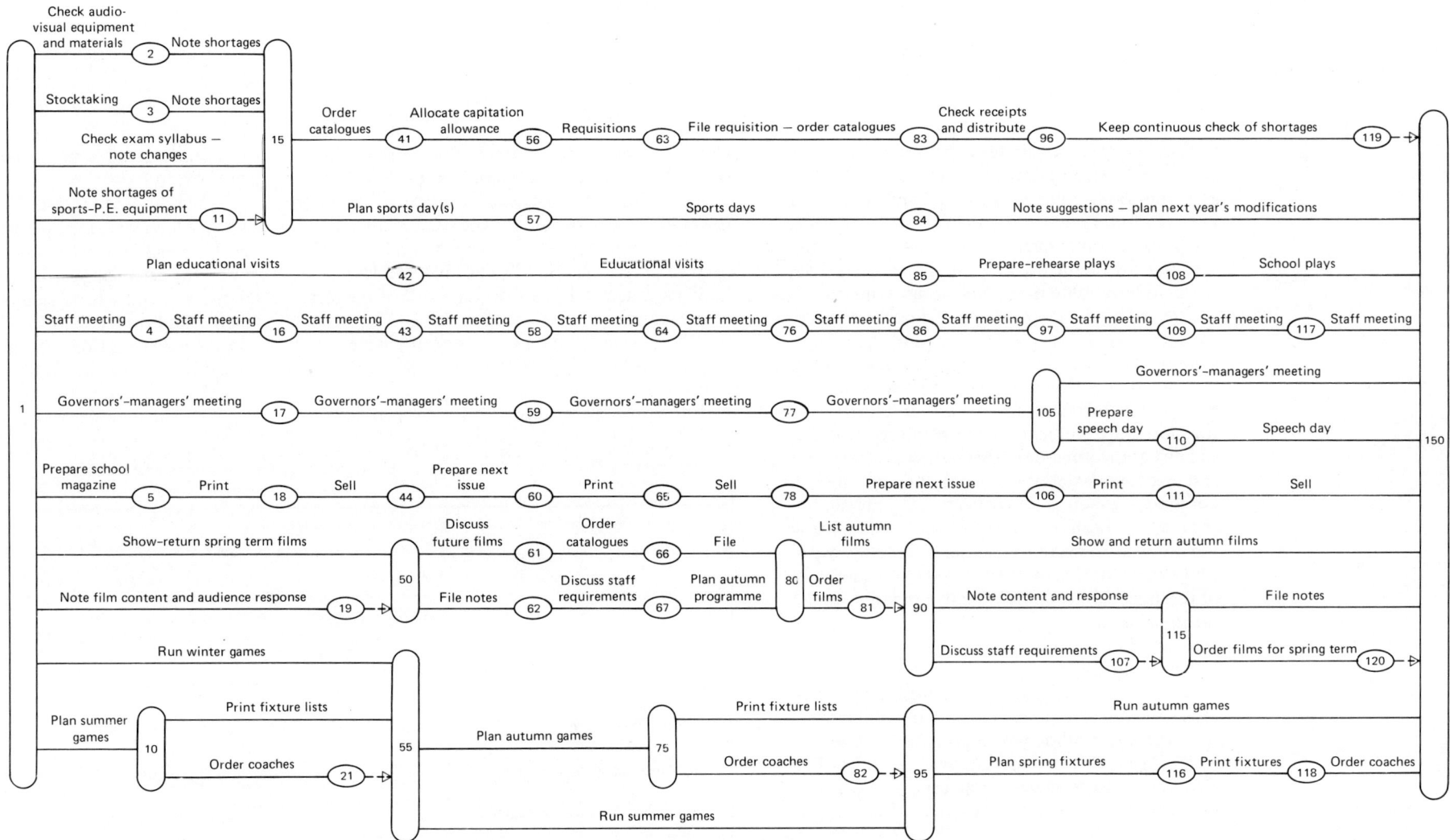

Fig. 54. *A simple network analysis of some of the annual events which take place in many schools.*

ordering, and ensuring that the correct stock is delivered, will be linked with the examination policy of the school. In many departments the books and equipment ordered will relate closely to examination requirements. To ensure that the ordering is done most effectively members of staff will have to be fully up-to-date. This might involve them in attendance at courses and conferences as well as wide selective reading.

The educational visits made by the school may be one-day visits to individual places of interest, such as castles, museums, factories, sporting events and art galleries, or they may be more extended. A network for a one-day visit is given on p. 94 and on pp. 97–100 there is a detailed network and account of a week's residential excursion involving visits to a variety of places of educational interest. Schools planning visits abroad might adapt this network to their particular needs with advantage.

Not all schools have their own magazine, but those which do not aspire to a full magazine often have an information sheet which is sent out to parents to keep them informed about the various activities in the school. These help to foster greater co-operation between parents and school. Whether the school produces a magazine, or a news sheet, there are a certain number of irreducible tasks to be performed. In producing a news sheet only one member of the staff may be involved, but that member must have an overall plan in mind. If a more ambitious magazine is to be produced either termly or yearly then it might be worth producing a network analysis of the activities involved to help with organisation, timing and delegation.

Many schools make full use of audio-visual aids and if this is to be effective a considerable amount of planning is necessary. This is dealt with in some detail. A section in this chapter (p. 107) is concerned with the choice of audio-visual equipment. Chapter 2 uses the showing of a film to illustrate the technique of network analysis and a section of chapter 4 (pp. 37–8) also deals with audio-visual aids. Many schools have their own television or film

units, or they may use facilities borrowed from colleges of education or teachers' centres. In making films and programmes there are a large number of activities which may involve several individuals or groups. A network of the production of an educational television programme is included as an example of the planning involved (p. 109).

School concerts usually do not take up a year of school time but some members of staff may well be visiting libraries, attending courses, and making draft plans for productions later in the year, and gradually other members of staff and pupils become involved as the year progresses. Some of the aspects of organising a school concert are illustrated in a network on p. 93. Speech day is treated similarly (figure 55).

The programme for sport varies immensely from school to school but the arrangements for individual matches of various sorts usually run to a well tried and familiar routine. For a sports day, however, more detailed planning might have value, and a network of the planning and organisation of a typical sports day is included on p. 104.

Since the design and production of a teaching programme can be complex and may involve a number of members of staff with different skills in different activities, a network for such a production is printed on p. 105.

As for the overall planning of the school year, if a headmaster uses the network or an adaptation of it, he could also use more detailed networks of particular aspects of the work of the school, and perhaps issue them to members of staff as an example of the sort of planning which they could then adapt to suit their own needs.

Finally it seemed worth showing how these planning techniques could be applied to students' teaching practice, for the benefit of students, their tutors and the schools they visit (p. 110).

School speech day

The network diagram gives an idea of the activities

which might be associated with a school speech day. In some schools speech day is the highlight of the school year, while in others it is one of a number of events and no special provision is made beyond the minimum (figure 55).

There might have to be provision for accommodation for the invited speaker and his wife if they have to travel far, but this would not be very common. Some schools which have only a small hall might find it necessary to hire a hall. This kind of thing would need to be added to the network.

Once the series of activities for a school has been decided the next task is to delegate responsibility. For example:

Headmaster: guest list, invitation of speaker and liaison with governors
Deputy headmaster: prizes, programme, dealing with cups, certificates and amplifier.
Music specialist: choice of items for choir and rehearsals
Domestic science specialist: provision of refreshments

Although the network describes a range of activities under the heading 'speech day' it is obvious that with very slight modifications the network can be readily adapted to any similar event. These may range from talks given at parent–teacher meetings, bring-and-buy sales, fetes, jumble sales, open days when parents are invited to visit the school, dances, and so on. If these events are to be occasions which pass without a hitch, prior planning is needed, and the larger the school the greater the need for care in planning. The network can be adapted, or used as the basis for a new one to meet the particular requirements of any school event.

School concert

School concerts vary from more or less informal events in which each class of pupils presents a short sketch, dancing, or display of gymnastics, to joint ventures of a more ambitious nature in which the whole school becomes involved. It is the latter kind which calls for good planning and organisation if it is to be a success and run smoothly (figure 56).

The network shows the activities which would be common to most school concerts and does not attempt to make provision for every possible eventuality. If used the network would need careful examination and additions and modifications to suit it to the particular circumstances and requirements. Some of the activities might need to be broken down into smaller activities. For example before the staff meeting the teacher who chooses the plays, or music, would need to survey the possibilities and make a short list for presentation at the staff meeting. This might then be further shortened and a final decision made at a later meet-

ing. The member of staff who made the first selection would want to be aware of the needs for the whole year, and would visit libraries and seek advice and suggestions from as wide a field as possible.

With the network as a basic plan the delegation of tasks can be considered. For example:

Headmaster: overall supervision, invitation of guests. This latter might be in consultation with managers or governors

Deputy headmaster: posters, tickets, seating and the hall.

If the school hall is not large enough an outside hall may need to be hired, and used for rehearsals and the actual concert. It is likely that as the time of the concert draws near a timetable will have to be made for the use of the school hall for

rehearsals and the normal hall timetable suspended.

Domestic science: refreshments, costume production.

Art department: designs for tickets, posters, programmes and costumes, and design and production of scenery. The latter might be in co-operation with the woodwork or metalwork department.

Science department: lighting and effects.

No attempt has been made to identify the critical path; conditions vary so much from school to school that this would be of little value. However, when a school has decided on an overall plan and draws a network the critical path could usefully be identified.

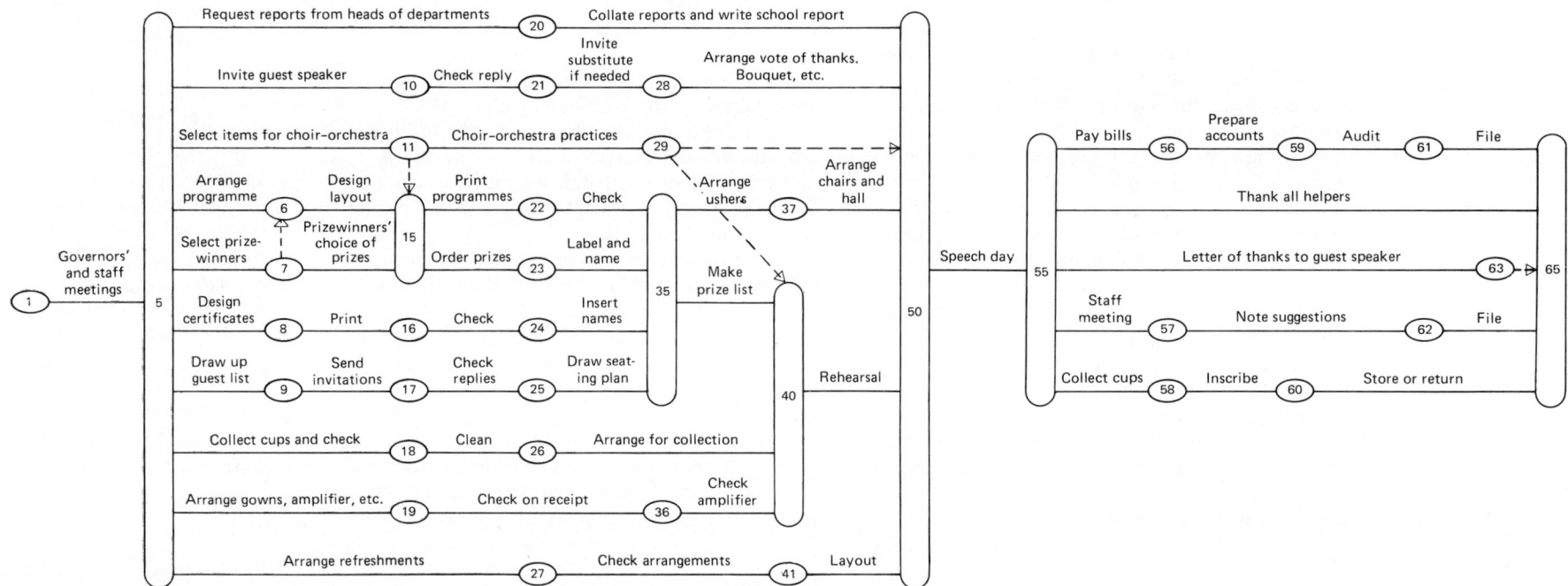

Fig. 55. *A network analysis of some of the activities involved in a typical school speech day.*

Educational visits

If visits are to have maximum value some care must be taken in planning the work involved. There is little value in a single unrelated visit to a place of interest. Ideally any visit should be thoroughly integrated into a scheme of work. It is at the first staff meetings that the overall scheme of work will be examined. Once it has been decided what sort of visits fit into the school plan of work, teachers can set about finding out which of the options open to them would be most worthwhile. As coach travel is now fairly easy and comparatively cheap any place of interest within a twenty mile radius of the school is a possibility. It is at the planning stage too that a study should be made of the results of previous visits. Over a period of years the staff will have gathered a general fund of knowledge about local places of interest. The local historical or geographical societies will also provide information. Once the plan for the year is drawn up and agreed a detailed plan such as the network analysis in figure 57 can be used to advantage.

Any visits to places of educational interest can use this plan with minor modifications.

Once a plan has been agreed different members of staff can take responsibility for a line of activities. One member of staff could take responsibility for transport and insurance. All would be involved in the design, production and use of any teaching materials which are required. Very often handouts will be helpful and these can be designed by different members of staff. The art specialists might be willing to help with the design. Any audio-visual materials can be ordered and used at an appropriate time.

Most schools will not only be able to use the resources within the school but will also be able to call on outside sources. Local education officers and advisers will often give a great deal of help by lending different types of equipment. The local library service will often lend books and perhaps

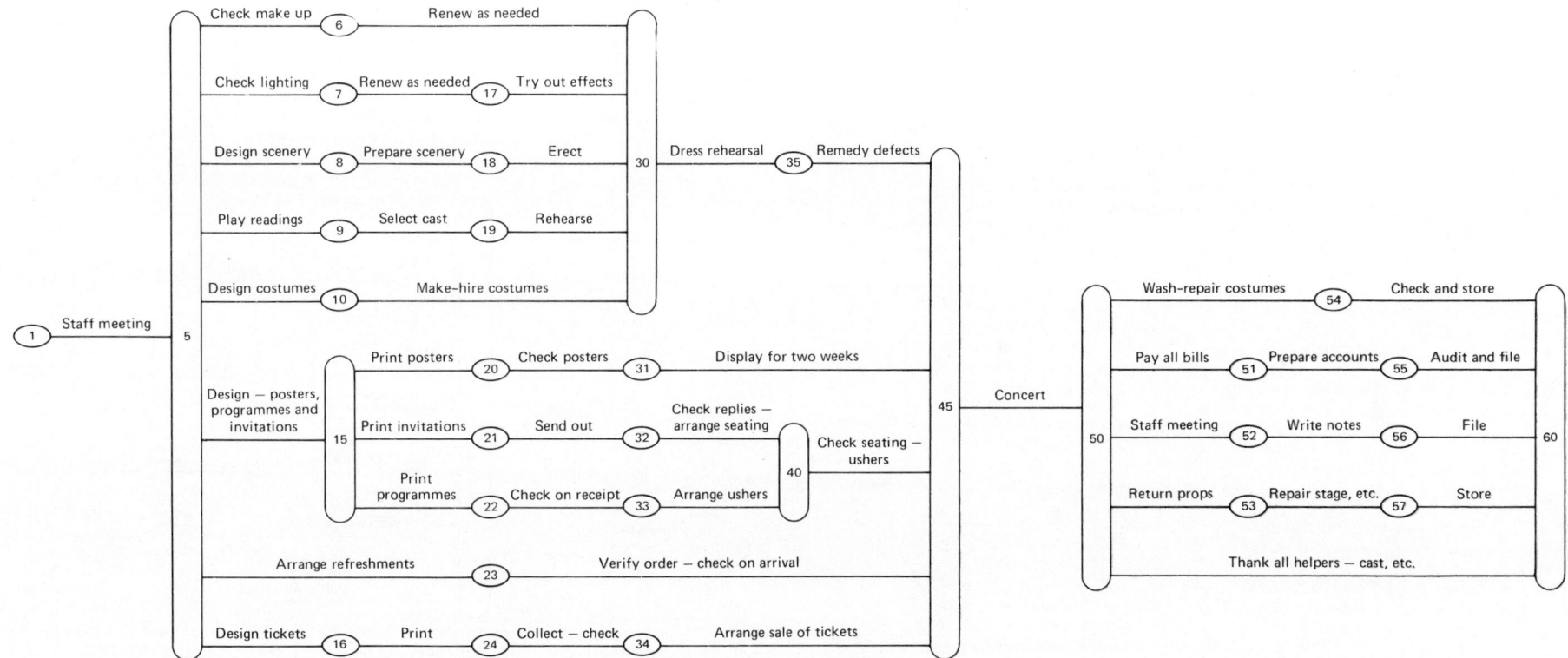

Fig. 56. *A network analysis of some of the activities involved in the production of a typical school concert.*

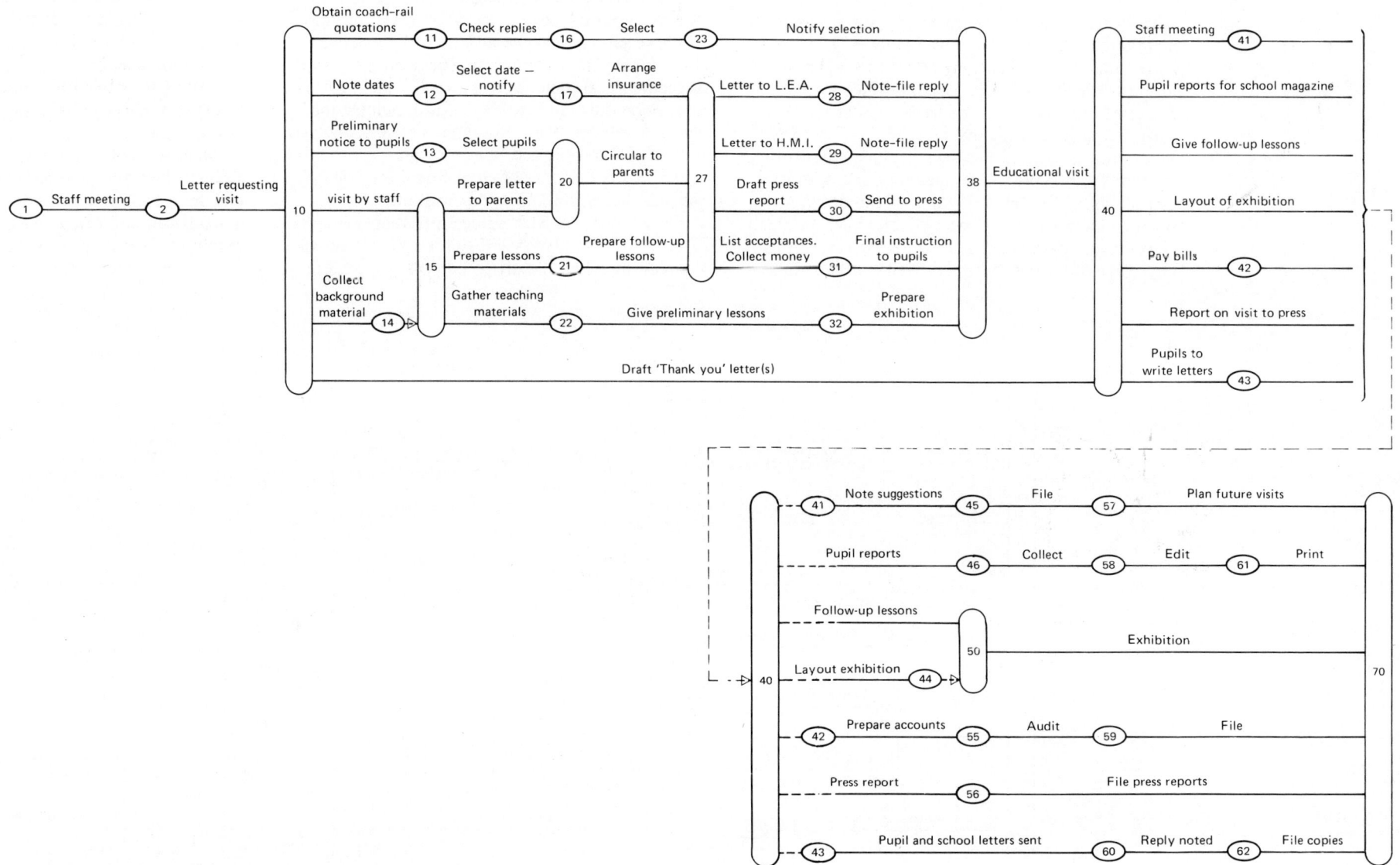

Fig. 57. *A network analysis of some of the activities involved in a short educational visit to a place of interest.*

films, film strips, tape recordings and other materials.

It is often necessary to try to obtain the help and support of parents. Even though expenses are kept to a minimum, they must be fully persuaded of the value of such visits. To this end some care should be taken to display the work done by the pupils in an attractive manner after a visit and parents should be invited to an informal evening for this purpose. Here they can not only see the pupils' work, but may also be able to see photographs and ciné film of their children. This is a very valuable exercise and is much appreciated by parents, particularly in the case of longer and more elaborate visits. One important aspect of the educational visit is the training which can be given in an informal way in correct behaviour so that the minimum of inconvenience is caused. Polite letters of thanks are usually much appreciated when sent to those who have helped to make the visit a success.

It is necessary to plan for the accounting of the money involved. It is only prudent to ensure that not only is the money correctly disbursed, but also that it can be shown to be correctly disbursed by the production of properly audited accounts.

Longer educational visits

Any school visit requires a certain amount of planning if it is to give the maximum benefit. The activities in which the teacher is involved for a one-day educational visit are fairly easy to keep in mind, but when an extended visit is under discussion the planning involved can be quite considerable.

Before describing in some detail the planning and organisation involved in a particular educational visit there are some general points which should always be taken into consideration.

Danger. Some types of educational excursion may involve possible danger to pupils, and teachers should be aware of this and make plans to minimise any possible risk. Events which involve hill walking, or trips over or near mountains, sailing, canoeing,

or walks near running or still water have special hazards. Pupils must be informed clearly what they may and may not do. Staff who are used to the conditions and have experience must always be on these trips. They should have training in rescue and resuscitation. First-aid kits should be taken and medical help sought in deciding what is necessary in a first-aid kit. Pupils should not be taken if they do not have the correct type of clothing. For walking they need stout, laced boots or shoes. Their outerwear should be warm and waterproof.

If there is any possibility of getting lost or delayed, the local rescue team leader, or the police should be notified of the time when arrival at the finishing point can be reasonably expected, and of the planned route. In the event of any difficulty rescue teams can more easily find the party.

It is essential that there is effective discipline. In school if pupils do not do as they are told there are times when this may not be too serious. When out on an educational trip lack of discipline may be fatal. Pupils must at all times be told quite clearly what is expected of them, and what to do in an emergency. On most visits it is wise to have along a member of staff who has already been along the route of the journey, who has timed the walks and surveyed possible hazards.

I was involved in planning and leading a week's visit to the Isle of Wight. Forty pupils, ten and eleven year olds, took part. Three other members of staff accompanied the party. Having all the activities set out in some detail may be helpful to other schools planning similar visits. It is much easier to alter a plan to suit individual needs, than to start with a blank sheet. It will help if any snags are mentioned so that these particular pitfalls can be avoided.

For the visit described, accommodation was in an hotel. This was thought better for younger pupils. For older pupils youth hostel accommodation might be used. Although it is more spartan, there is value in working as a team in such activities as cooking and cleaning.

The planning of visits abroad varies depending upon whether the pupils travel as a party, and then live with families, or whether they travel and stay as a party. It will be necessary to have members of the staff who are fluent in the language, and it helps if the pupils have some basic teaching in it as well. Pupils should be instructed that the food may be different and be prepared to try anything at least once. Teachers must prepare for passports, visas and currency to be ready in time. Regulations vary from time to time and teachers should have up-to-date information (figure 58, pp. 97–100).

Pre-planning. Any information gained from previous educational excursions made by the school should be studied, especially if they have been to the same area. Members of staff who have made such previous journeys should be consulted and their suggestions noted.

There should be a preliminary staff meeting for all those who may be involved at some time. This ought to be a term before the projected date of departure, or about two months. If the visit is to be of any educational value there is a lot of preliminary work to be done by both staff and pupils.

The staff meeting should have on its agenda the following:

1 *Whether to have the visit at all.* The results of previous visits may have caused members of staff to decide that they would not go on another visit. There may have been a series of mishaps, involving bad planning, poor accommodation, bad behaviour by pupils, accidents, poor educational results, lack of time available, poor staff relations, or just lack of interest — any of these might cause an educational visit to be a misery. In which case the decision might be not to have a visit.

If the decision is to go ahead then the rest of the agenda can follow.

2 *The place to visit.* This might be the same as last year. If a new place is under discussion then

some of the considerations to be borne in mind might be:

(a) The distance from the school. It should not involve too long a journey for young pupils. They can become tired, bored and troublesome after a long journey.

(b) The place should be a change from the school's surroundings, and offer variety. Ideally there should be a balance of the industrial and the rural. There should be a variety of places of historical interest, and if possible a variety of high and low land, and inland and coastal scenery.

(c) The accommodation available should be comfortable and should include a space which can be used for the pupils to assemble to hear talks and to do any work required. The food should be of a reasonable standard and, for preference, child centred. Children today are accustomed to some innovation but they are fundamentally conservative in their choice of food. While one would not advocate fish and chips and baked beans with every meal, any chef who tries to introduce too many new dishes, even though they may be excellent, will find much food left on the plates.

3 Dates. When a suitable place has been chosen, the dates should be agreed. This may bear on the next item on the agenda, the staff available. Generally schools try to avoid the popular months. Care must be taken when deciding to go early or late, to allow for cold and wet weather. This means ensuring that the pupils have the clothes for such weather, and making contingency plans so that bad weather cannot ruin the visit.

4 Staff. One needs a team who will knit together into a closely co-operating group. Close involvement by all members of the team in all the planning, and full discussion of all the work involved can be a big help in knitting together such a team.

Although one need not do this too formally, it is necessary to split up the duties and responsibilities. Members of the team will have special interests or skills. Full use should be made of these in allocating duties. Our team split up conveniently, as we had one teacher interested in history and buildings; one in plants and animals; one in geology and geography; and one in industry and planning.

The particular interests of the staff available will have a bearing on the next item on the agenda. It is obviously no good arranging visits to places which are not of interest to the staff who are going.

5 Programme. The draft programme should be fully discussed. Full account should be taken of any previous experience of visits to the area, visits made by staff members, and any notes and informal information. If any member of the team has a particular interest, a visit taking this into account may be of special value. The programme should be as flexible as possible.

Care should be taken not to involve the pupils in too long, or too strenuous days. We found that a six mile walk was about as much as the ten and eleven year old pupils could manage. We planned to alternate fairly quiet days and more strenuous days involving more walking. This proved to be about right. In addition the days were planned so that after dinner the pupils were quietly engaged in writing up the day's notes or finishing sketches made during visits to places of interest. This worked well as most were healthily tired from the day's activities, and welcomed a quiet time before going to bed.

A decision must be made whether to use a hired coach for visits, or public transport. Out of the holiday season public transport can be used, but one is unpopular using it during the busy periods when workers are off to work or returning. If the weather looks unsettled it is a help to have a coach which can be used as shelter.

In trying to time expeditions it is as well to underestimate the walking speed of the children.

Some of our pupils, we found, walked very slowly. This was partly due to unsuitable footwear. It was also due to their interest in all about them. You cannot hurry pupils past a spot where a snake has been seen! While one would not wish to hasten children too quickly, stops to observe are time-consuming. I think we managed about two miles an hour at best. But one should not allow a timetable to dictate too much.

After the initial discussions and draft planning there follow a series of closely interlinked activities.

Parents are involved. They need to know in good time the sort of clothes to provide. This may be be easy. Much of the clothing worn by pupil quate for normal wear when they are ne from shelter, but is far from adequate to be out of doors for a day. Many children's clothing are 'showerproof' and this is not sufficient for a possible whole day in the rain. While contingency plans will allow changes of programme when foul weather is certain, one can be caught in the country far from shelter. If at all possible pupils should have waterproof clothing. In the same way footwear suitable for day-to-day use, may not be suitable for an extended walk in the country. After full discussion a list should be prepared and sent out to parents of pupils likely to go on the visit. Clothing needs can be emphasised at a parents' evening. Here the draft plan can be discussed in detail. Also, in case of accident or illness it is essential to know of any allergies to normal medical treatment. Some pupils have to take anti-asthma pills or similar medication. Parents must be asked for this kind of information. Some children, for one reason or another, will drop out of the visit, so a reserve list should be drawn up.

If the area has not been visited before, letters have to be written to hotels and the most suitable chosen. It is best if someone has had personal experience of the hotel selected, or if it is chosen on the basis of a recommendation.

The party needs to be split up into groups,

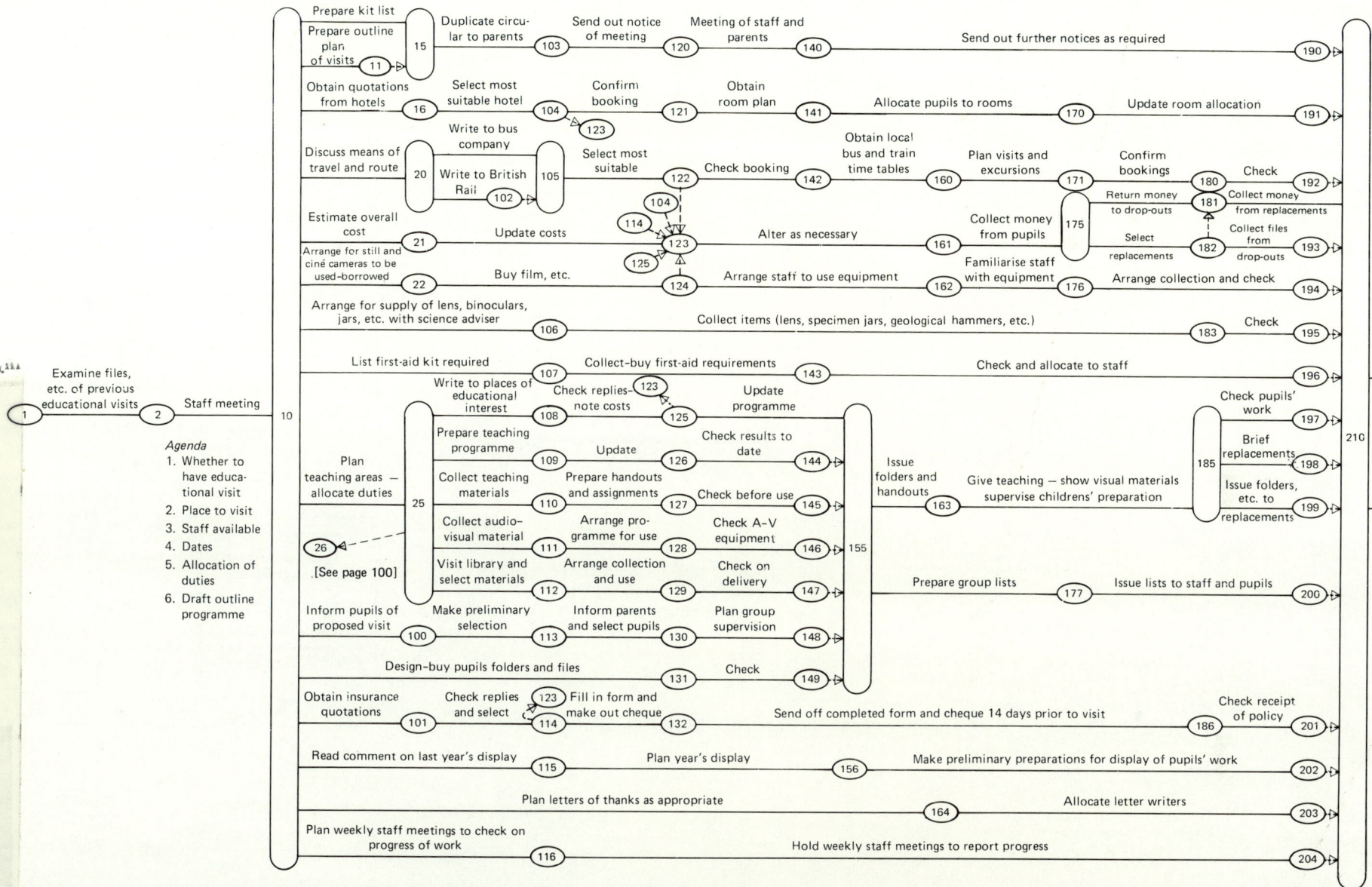

Fig. 58 (part 1). *A network analysis of an educational visit to the Isle of Wight for the period of one week.*

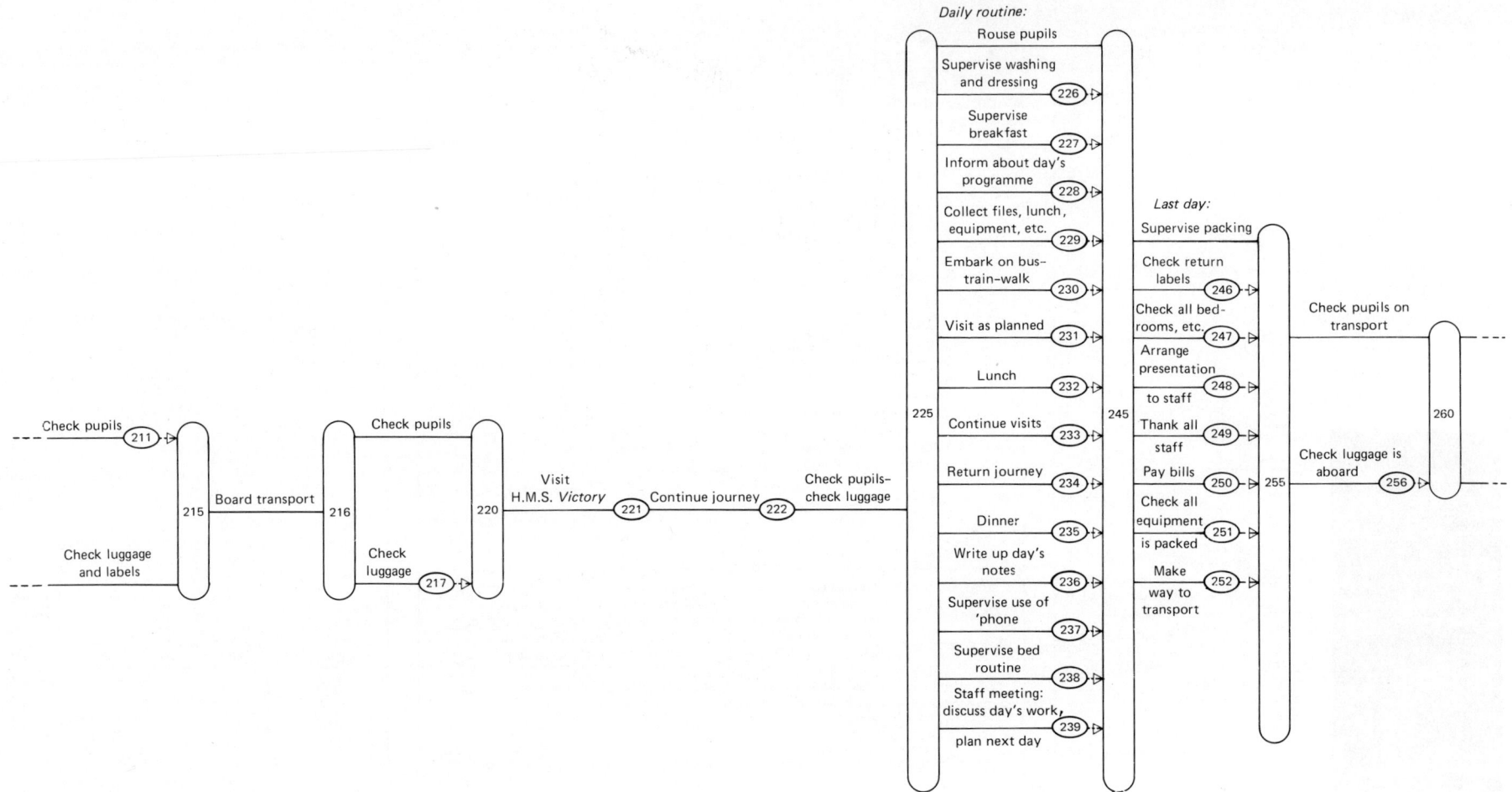

Daily routine:
Rouse pupils
Supervise washing and dressing (226)
Supervise breakfast (227)
Inform about day's programme (228)
Collect files, lunch, equipment, etc. (229)
Embark on bus–train–walk (230)
Visit as planned (231)
Lunch (232)
Continue visits (233)
Return journey (234)
Dinner (235)
Write up day's notes (236)
Supervise use of 'phone (237)
Supervise bed routine (238)
Staff meeting: discuss day's work, plan next day (239)

Last day:
Supervise packing
Check return labels (246)
Check all bed-rooms, etc. (247)
Arrange presentation to staff (248)
Thank all staff (249)
Pay bills (250)
Check all equipment is packed (251)
Make way to transport (252)

Check pupils on transport
Check luggage is aboard (256)

Check pupils (211)
Check luggage and labels
Board transport
Check pupils
Check luggage (217)
Visit H.M.S. *Victory* (221)
Continue journey (222)
Check pupils–check luggage

215 216 220 225 245 255 260

Fig. 58 (part 2). *A network analysis of the journey, daily routine and plans for return.*

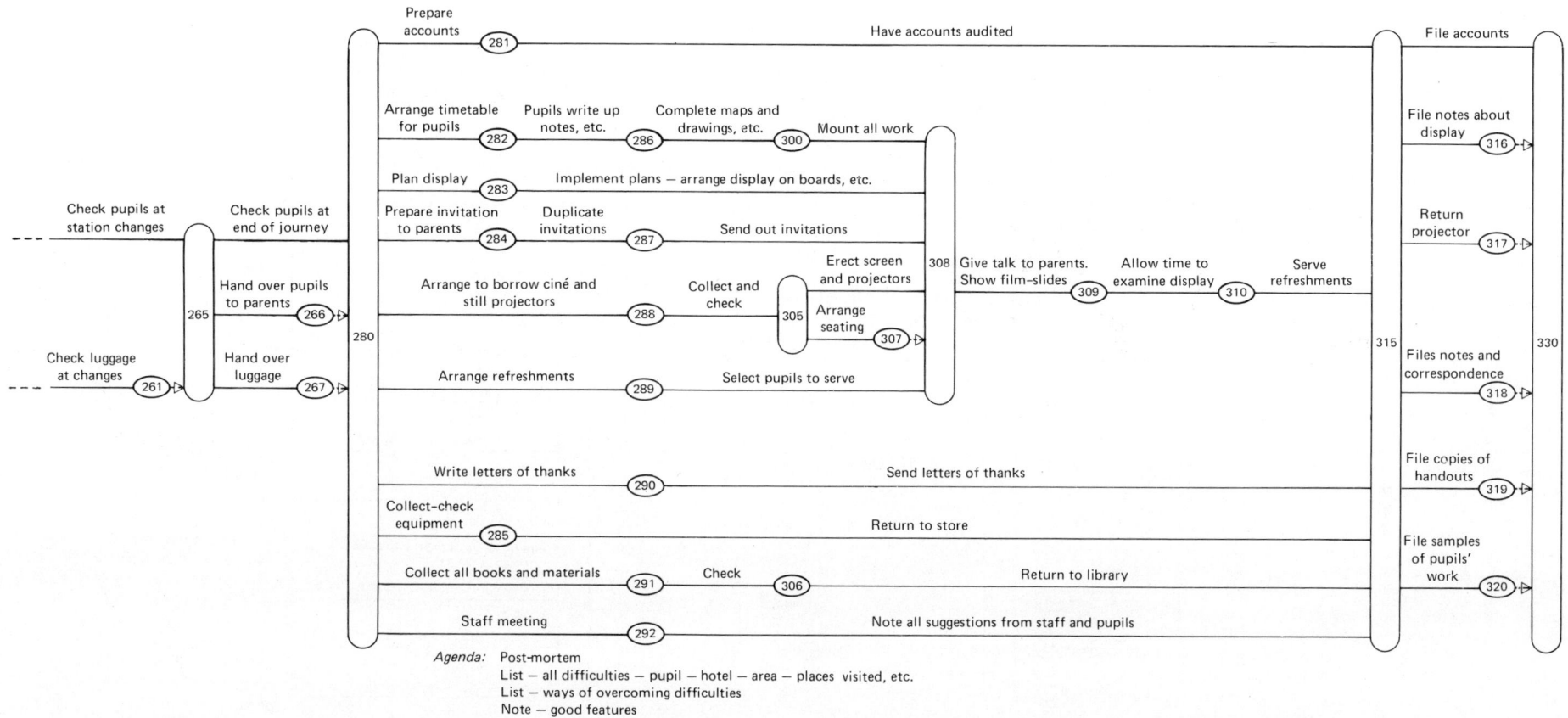

Fig. 58 (part 3). *A network analysis of post-educational visit activities.*

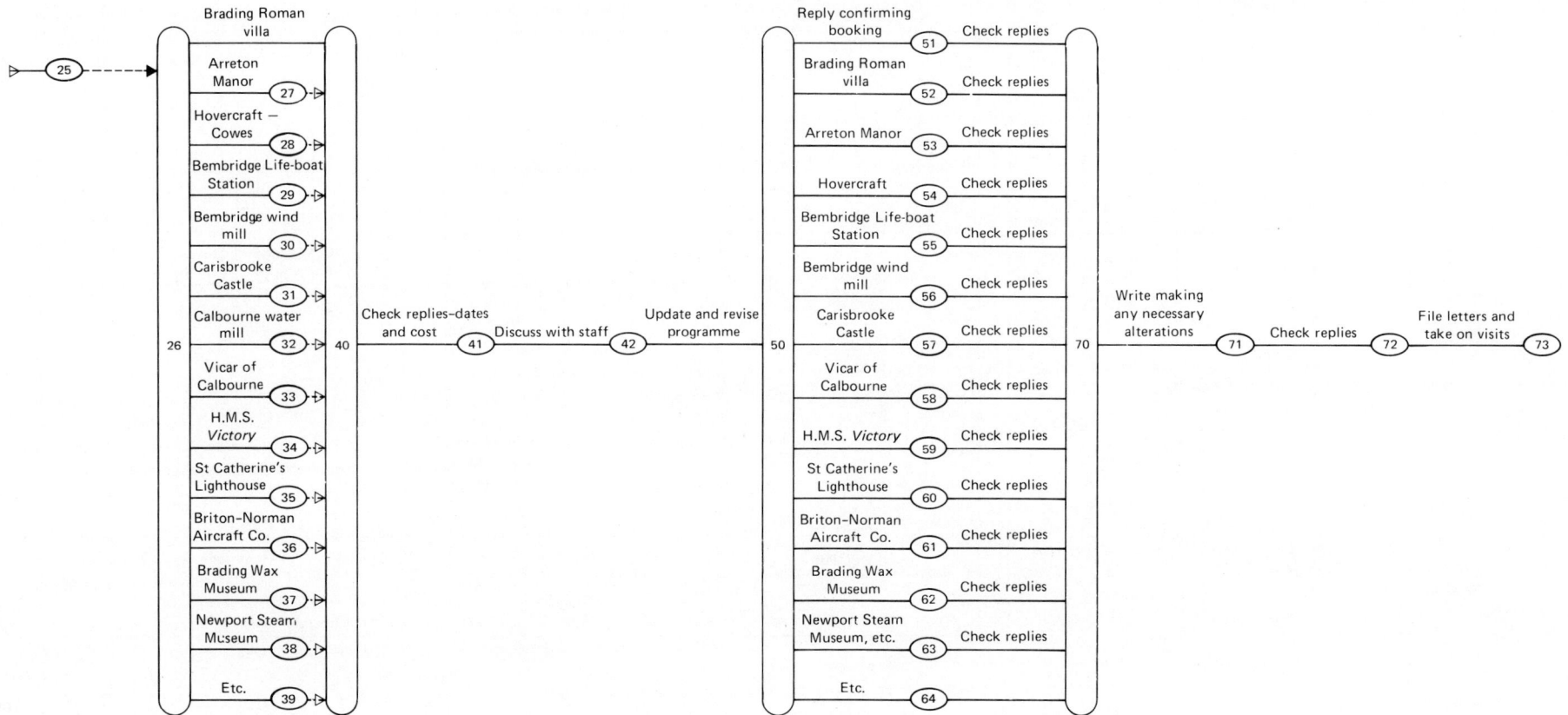

Fig. 58 (part 4). *A sub-network analysis of some of the correspondence involved in an educational visit to the Isle of Wight.*

each group being the responsibility of one member of staff. In our case since we had forty pupils and four members of staff, each had special responsibility for ten pupils. We did not insist that pupils stayed in groups except where extra care was called for in supervision, on the ferry or during changes of train. It saved a lot of time at roll-calls to have a limited number to check.

I thought it best to allow pupils to choose bedroom partners, and this worked out very well. There were some last minute requests for changes from some girls who had fallen out with their friends, but the problems involved in such changes were pointed out and they stayed put. In the same way I asked the children which teacher they wished to have as group leader. This was taking rather a chance as they might have all chosen one member of staff. In the event the groups worked out evenly. I noted that pupils picked as group leaders the teachers in whose class they had been, or the teacher in whose class they were, possibly taking the view, better a devil you know than one you do not!

Transport needs some discussion. Coach travel is at the time of writing cheaper by about half than rail travel. It is convenient in that pupils can be picked up with their luggage at a convenient point. But travel by rail is usually quicker. If the journey is of any distance this can be an important factor. Long journeys are tedious for most pupils after the initial excitement has worn off. There is a limit to the ways in which pupils can be engaged in suitable games, or in observation of the scenery. When quotations of cost have been studied and the method of transport has been selected, bookings will have to be made. This should be done well in advance.

In addition to travelling there and back, transport will be needed during the stay. Many of the decisions about this cannot be made until the draft programme becomes firmer. Then suitable transport should be selected. Usually local rail transport is limited, so the choice comes down to public bus services or a private coach. If a private coach is to be used booking in good time is essential as even in the slack season many coach firms are fully occupied transporting pupils to schools and workers to factories.

There are good reasons for arranging for the pupils to walk when this will not put an undue strain on them. Many pupils today seldom enjoy the delights of walking in the countryside. For children interested in plants and animal life such walks have great value.

At an early date it is necessary to make an overall estimate of the total cost of the visit. The cost of transport and hotel can be fairly quickly established. For the rest a rough estimate can be arrived at, and a proportion added for pocket money. This estimate will need to be included in the first circular to parents so that they can make provision. We allowed the pupils £2.00 pocket money, and this was in the charge of the group leaders who issued an agreed amount to meet the needs of each day. This worked out quite well and all the pupils had sufficient for post-cards, souvenirs, bus travel and for little presents to carry home. They also all made a small contribution towards presents for the staff of the hotel. It is advisable for teachers to take charge of pocket money and issue it as needed, rather than allow the pupils to keep their own; otherwise some will have none at the end of the first day.

At the start of the project provision should be made for a system of accounting. Care in planning and care in keeping accounts will prevent the embarassment which can be caused by a bill which arrives after it is thought all accounts have been settled. Whenever possible bills and receipts should be kept and all money spent should be listed and accounted for. It is often not only necessary to be scrupulously honest, but for this to be seen to be so. I have found to my cost that if I do not keep a careful record of money spent on these ventures, I end up out-of-pocket. Full accounts should be available for parents to inspect if they wish.

After educational visits it is good to be able to show parents the places visited by their children, and their children in action. It is worth having some photographs taken. These can be placed on display and add interest, or even amusement, to an opening evening. We also borrowed a ciné camera from our local educational technology centre.

In order to take full advantage of the various aspects of the places to be visited the team approached the local rural science staff and asked if they could borrow binoculars, geology hammers, specimen jars and lenses. These were a great help. The pupils were instructed that they should look, but not touch or remove from their habitat any plants or animals. A collection of detached seaweeds, shells and pebbles was made and examined at leisure in the evenings. Most were identified using the reference books taken with us. Notes and drawings were made of the collection. The pupils had been given sheets of drawings of the types of birds, plants and fossils they were expected to find. They were able to use these sheets for most of the specimens, and to turn to the reference books for those uncommon varieties they found. Full use should be made of any specialist departments of the local authority. Not only are they often prepared to lend items of equipment, but they will also help and advise in many other ways.

We have for some time had ready-packed first-aid kits in the school. These can be picked up and taken whenever a party goes out on a visit. These contain bandages, antiseptic cream, plasters and other simple first-aid items that may be required. The local medical officer of health would be pleased to supply a list of suitable contents to schools. We have so far not needed to use our kits except for the odd plaster or two, but they should always be included. The kits should be spread out among the staff involved in any excursion so that they are ready to hand at once.

As far as the teaching part of the venture is concerned it will quickly be ascertained which members of staff have special interests and they can be given responsibility for that particular area of the project. Then each teacher can plan his part of the programme. Suitable handouts can be designed and duplicated, and collections of books can be borrowed for the pupils to use for reference.

It is necessary to co-ordinate all the separate efforts into an overall scheme. In order not to upset the work of the school, as the pupils to go on our visit were from six different classes, we arranged for preparatory meetings to take place during lunch breaks. All reference materials, charts, maps and anything which needed to be displayed was kept in one classroom. Any pupil who was going on the visit was told that he could use this room at any spare moment, during the lunch periods, and at break times. The room was well used. In addition two or three times a week for a month prior to the visit the teachers took it in turns to talk to the pupils, to issue handouts, and to show slides and film strips covering particular aspects of the visit and to give the necessary background information.

The local education library service provided a stock of reference books, film strips, slides, charts and maps. We took a large number of reference books with us. In fact more than we needed. We should have concentrated on taking a few comprehensive reference books, rather than a lot of less comprehensive books.

Paper for drawing and notes can be a problem. Handled outdoors, often in windy conditions, paper soon becomes very much the worse for wear. To overcome this we used plastic folders. They keep papers clean and are also reasonably waterproof. We also purchased for the school some lightweight nylon haversacks and these were also invaluable. We have found in the past that the usual kitbag is not suitable for journeys as it keeps one hand occupied or, if worn over the shoulder, the string tends to cut after a while.

It is necessary to take out additional insurance when taking parties out from school. Rates vary considerably and it is worth getting a number of

quotations. The premium needs to be sent off in good time, usually at least fourteen days before departure.

After any extended visit some sort of display of work should be given, perhaps on an open day or a parents' evening. This is particularly necessary when parents have been involved in considerable expense. They need to be reassured that their child has benefited from the experience. This should be planned at an early stage. Work done by all the pupils should be on display, together with mounted photographs, maps and diagrams. Films and slides can be shown.

Finally it is important to ensure that all who help are given appropriate thanks. This should not only be done by members of staff but also by the pupils. Giving thanks is a courtesy which is often forgotten. We detailed one child to give thanks on behalf of the party at each of the places visited.

In order to keep everything in step it is wise to have regular staff meetings in addition to informal discussion. Any slip-up in plans can be dealt with in time. Comments about the pupils, their strength and weaknesses can be of value. Those who are falling behind in their preparation can be encouraged. Any pupil whose behaviour is bad can be warned to improve his ways.

Towards the end of the period of preparation most of the visits to places of interest will have been confirmed and checked and a day-to-day programme can be set out. There must be a certain amount of flexibility to take account of the unexpected, and to take account of very bad weather. Copies can be sent home to parents and should be given to the pupils so that they know what to expect, and what provision to make for each day's events.

With plans laid covering all the major areas of activity, there were a couple of last minute problems which could not be covered by any planning. One pupil had contact with chicken pox. His family doctor finally declared two days before we set off that he was now unlikely to have caught the

disease. In the event he produced spots after three days. A doctor was called in and confirmed chicken pox. The school was phoned and his father came and collected him. This sort of unforeseen happening meant that one member of the staff had to remain behind for a whole day to look after the pupil.

This raises the question of how much allowance to make for unforeseen emergencies. There is always pressure to have the minimum of staff to supervise and it is generally agreed that a ratio of one member of staff to ten pupils is reasonable. This ratio should not be cut. In planning care should be taken to ensure that there is 'slack time' which can be used for the unforeseen event which causes delay.

Another pupil was called into hospital at the last moment to have a long delayed operation. Two other pupils had to drop out at a late stage as their father was suddenly put on short time at a local factory. In each case there was a replacement standing by, who had been doing the preparatory work, and had attended all briefings, who could step in. Money had to be refunded to those not going, and collected from those who were.

Duplicated sheets were produced and handed out to the children. They contained a range of information and assignments:

1   Weather chart for them to make their own entries.
2   General information about the visit.
3   Day-to-day list of activities, places to be visited, and the activities for pupils to engage in.
4   Outline drawings of common seaweeds for pupils to identify and colour.
5   Outline drawings of common seaside birds' footprints, and fossils to help pupils identify any found.
6   Outline map of Isle of Wight. Three copies for each pupil. On one pupils to fill in physical features of the island, on another under-

7   Notes about the seashore to give guidance to pupils.

8   Notes and drawings about wind-mills to be found — post, smock and tower.

9   Notes and drawings about water-mills.

10   Notes about forms of architecture to be found on the island.

11   Notes and drawings about common seashells to be found, identified and coloured by pupils.

12   General notes giving details about church architecture and what to look for in a church.

13   Notes about the journey and the places passed through.

A school sports day

It is obvious that some of the activities involved can take place at the same time as others. For example, on the afternoon the track is being prepared by the ground staff, the PE specialists could be conducting the long-jump and high-jump heats, whilst the deputy headmaster might be arranging the hire of amplification equipment for the announcer. This situation would be as shown in figure 59.

If it is decided that printed programmes are to be produced, arrangements have to be included in the network. This section of the work might appear as shown in figure 60, node 27 being the point

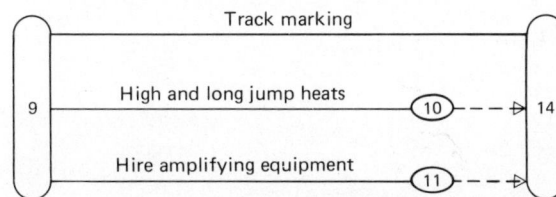

Fig. 59. *A section of a network analysis of a school sports day showing parallel activities.*

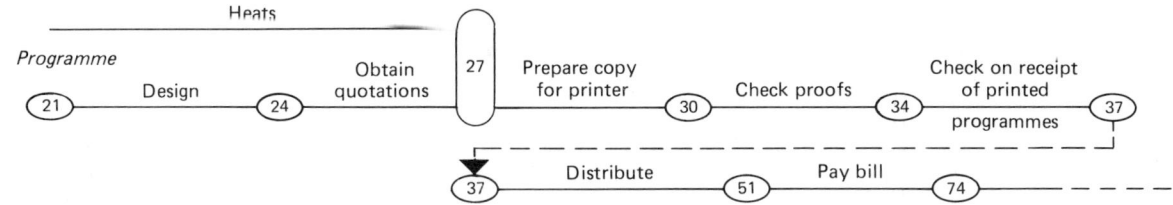

Fig. 60. *A sub-section of a network analysis of a school sports day.*

at which the results of the heats (the names of the finalists) must be available for the printer.

The complete network would, of course, need to be adapted to suit a school's particular needs. It would then be worth identifying the critical path (figure 61, p. 104).

Network analysis of writing a teaching programme

Programmed learning can be of value in schools where any form of individualised teaching is used. The writing, testing and production of teaching programmes is a time-consuming process and care and forethought are needed before embarking on such a project. It is in the planning stage that network analysis may save time and resources.

From the network (figure 62) it can be seen that the first activity is initial planning. This might involve decisions about the pupils who are to be taught, 5–15; the overall aims of the project, 5–6; the subject matter to be taught, 5–7; the method of writing the programme, 5–8; an estimate of the cost, 5–9; how long the programme would be likely to be used, 5–10; and how many pupils would be likely to use the programme, 5–11.

When the writing group have fully weighed up these considerations it is possible for individuals to go more deeply into each separate step. It might be useful at this stage to draft a flow chart, or a network analysis if it is a complex programme.

One aspect which is not often mentioned, as it cannot be planned so obviously into a programme, is pupil motivation. It is a vital consideration if

the programme is to succeed, and must be taken into account. If a pupil is persuaded of the value of a programme he will be motivated to start with some enthusiasm. If it enables him to succeed in the early stages this too will be motivating. Thus the programme will need to be presented at the right pace and at the right level of understanding — it will need to start at the right stage. Also the choice of media will have to be right.

The writers will have to consider (5–15) who the programme is for. Will it be aimed at all pupils of a certain age, or retarded pupils, or those taking particular examinations, or high ability pupils? They must decide what the pupil will be able to do when the programme is successfully completed and at what level of skill.

The consideration of objectives also requires a statement of the conditions under which the success or failure of the pupil will be judged. What questions will be set? How many? What proportion of correct answers will be regarded as a pass?

With the overall content of the programme decided, the next step is to break down this mass of teaching into increasingly small portions which can then be taught item by item. Programme writers call these steps successively sub-objectives, teaching points, and operants. The operants are the basic units of teaching upon which the whole programme is built. The size of these will depend upon the pupils. Older or high ability pupils will be able to cope with quite large operants, while younger or low ability pupils will need to have very small operants indeed.

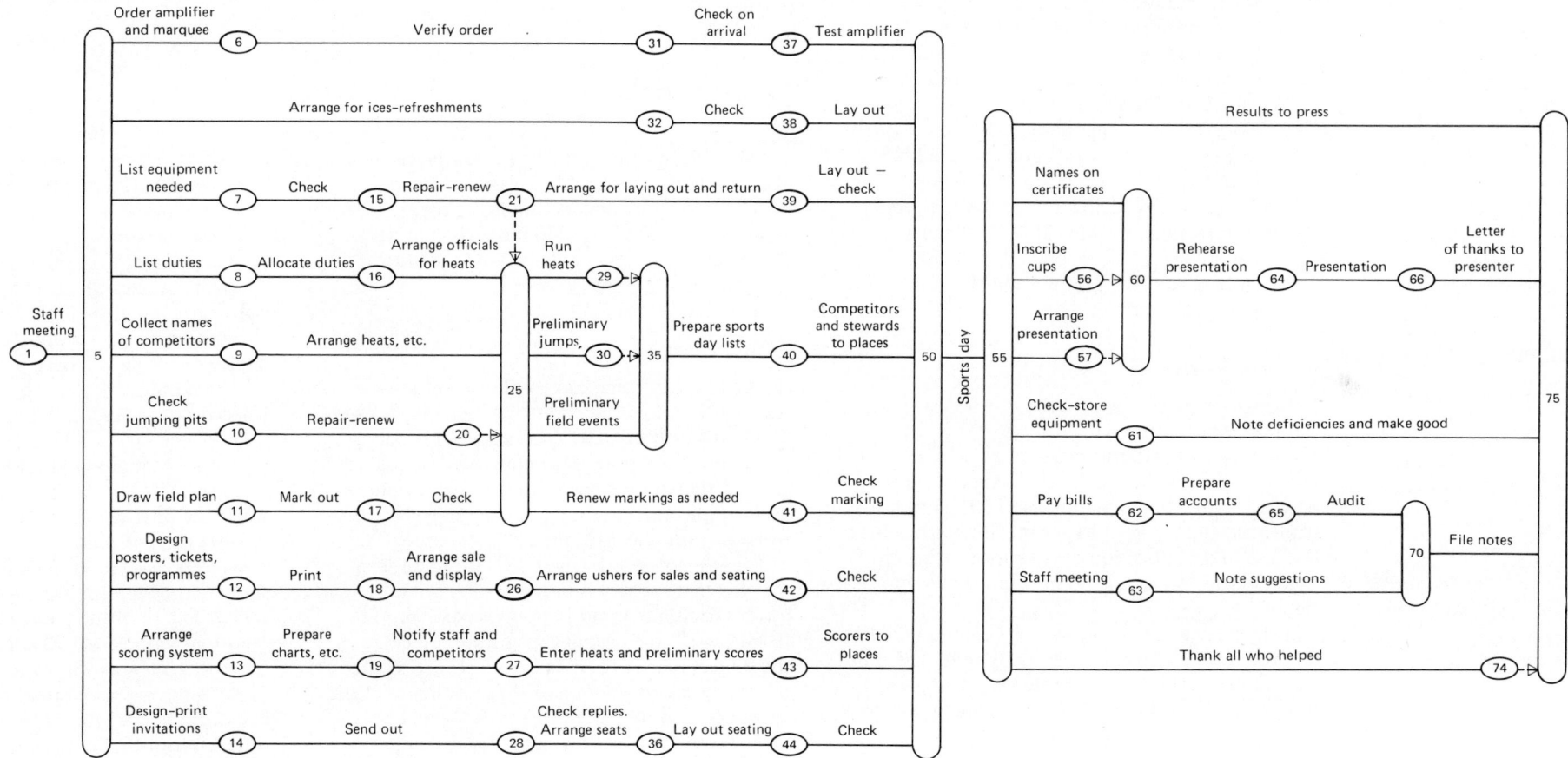

Fig. 61. *A complete network analysis of some of the activities involved in a typical school sports or athletics day.*

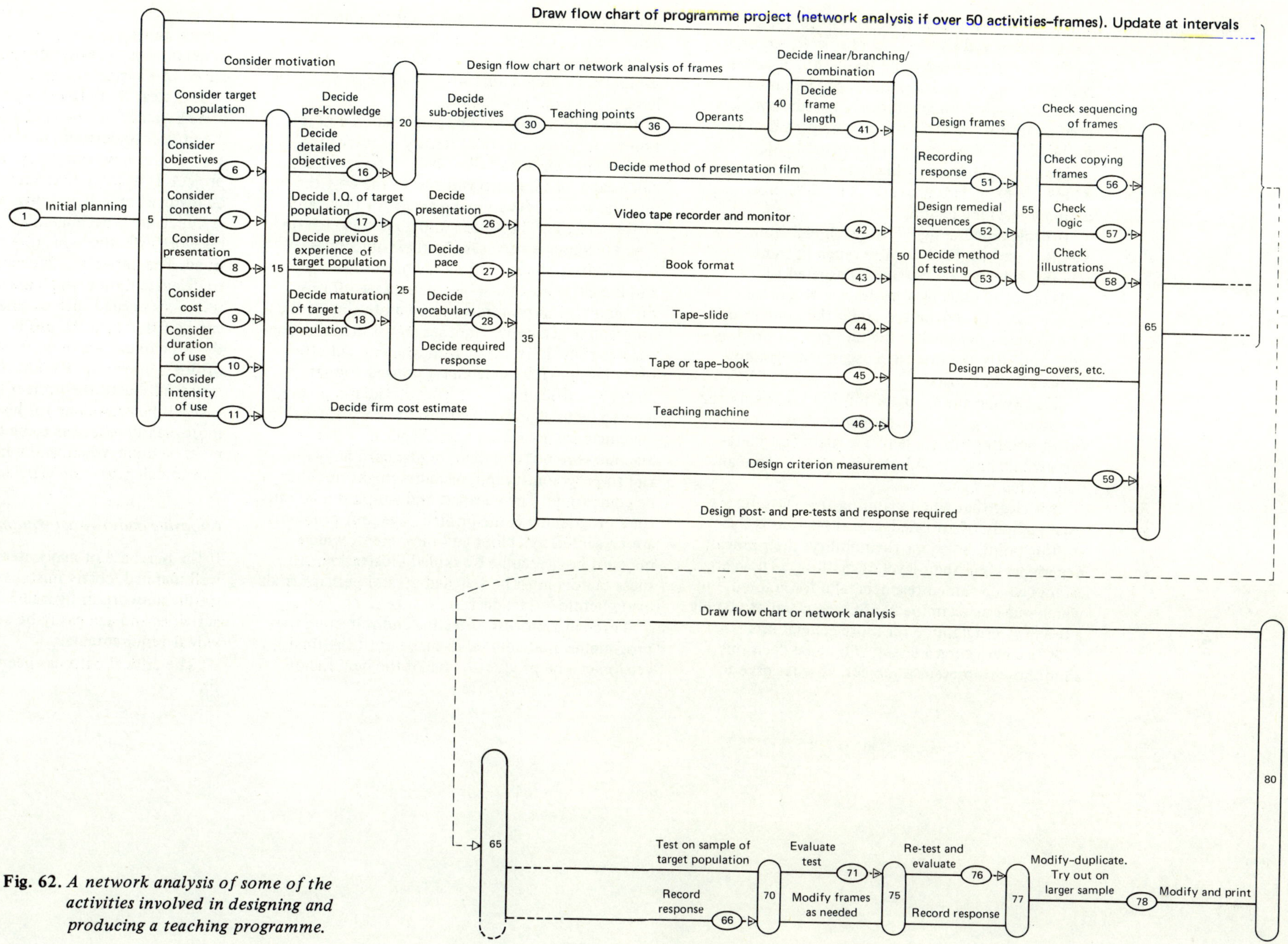

Draw flow chart of programme project (network analysis if over 50 activities–frames). Update at intervals

Consider motivation

Consider target population

Design flow chart or network analysis of frames

Decide linear/branching/combination

Consider objectives 6

Decide pre-knowledge 20

Decide sub-objectives 30

Teaching points 36

Operants 40

Decide frame length 41

Design frames

Check sequencing of frames

Decide detailed objectives 16

Consider content 7

Consider presentation 8

Decide method of presentation film

Recording response 51

Check copying frames 56

Decide I.Q. of target population 17

Decide presentation 26

Video tape recorder and monitor 42

Design remedial sequences 52 55

Check logic 57

1 Initial planning 5

Decide previous experience of target population

Decide pace 27

Book format 43 50

Decide method of testing 53

Check illustrations 58

Consider cost 9 15

Consider duration of use 10

Decide maturation of target population 18 25

Decide vocabulary 28 35

Tape-slide 44

Consider intensity of use 11

Decide required response

Tape or tape-book 45

Design packaging–covers, etc. 65

Decide firm cost estimate

Teaching machine 46

Design criterion measurement 59

Design post- and pre-tests and response required

Draw flow chart or network analysis

65

Test on sample of target population

Evaluate test 71

Re-test and evaluate 76

Modify–duplicate. Try out on larger sample

80

Record response 66 70

Modify frames as needed 75

Record response 77

Modify and print 78

Fig. 62. *A network analysis of some of the activities involved in designing and producing a teaching programme.*

The next three activities (5-9, 5-10 and 5-11) are all linked and have to be considered together. Different factors need to be taken into account when costing the programme. A programme produced to teach decimal currency would have been used a lot, but for probably a short period of time in the United Kingdom. On the other hand a programme which is not likely to date would have a long life-expectancy, as for example a programme teaching reading skills.

A programme designed to teach high ability pupils might not be used very much in some schools, whereas a programme designed to teach skills involved in using a protractor would be widely used. In general the larger the number using a programme the smaller the unit cost of production. Similarly the intensive use of a programme reduces its unit costs.

It is beyond the scope of this book to study the production of a teaching programme. But it is worth pointing out some of the main considerations which have to be borne in mind when planning such a programme.

It is clear that when planning any teaching the existing body of knowledge of the pupils is the starting point. So too is their ability, their previous experiences and their level of maturity. These are factors which skilled teachers take for granted, but which must be examined with care when producing a teaching programme for more general use.

A teacher giving a lesson will make decisions about how to present a subject, at what pace to teach it and what vocabulary is appropriate. He will be continually watching the response of the pupils and adjust his teaching according to this feedback. In programme writing all these decisions have to be taken in advance.

Presentation will have to be considered in relation to the duration and intensity of use of the programme. Since any film, slide, tape recording, video-tape, or film strip will add to the cost there must be sound educational reasons for using any of these media. The writers will have to weigh up how far colour is necessary to their objectives. If it is essential, slides or film may have to be used and the extra cost must be compensated for by the increased impact, motivation and value of the programme generally. Another factor might be the need to show movement. If movement can effectively be shown by diagrams a printed format might be sufficient. If not, film or video-tape may have to be used and will add to the cost. If teaching machines are used, cost again might increase. Machines are not common, or standard anyway, and so programmes and machines might not always be compatible. The cheapest and simplest medium for a programme is the printed page, but problems are presented by colour and movement, which may not be overcome by skilled illustration. All these factors must be weighed up and perhaps trials made before a final decision is taken.

Through a process of testing and re-testing the programme gradually takes shape until the final version can be produced. One of the benefits of the programming technique is that all programmes are self-testing in effect. Each step can be examined in detail. This is really normal skilled teaching reduced to a systematic and formal process.

It is clear that effective planning is important. It is obviously necessary that all those engaged in the project understand not only what they have to do, but also how what they do fits into the overall project. It is likely that a number of teachers will be involved and, if audio-visual media are used, technicians. If each member of the team is to do his work effectively his rôle must be clearly defined. The network will enable the project leader to delegate duties which are clearly set out. The resources needed such as ciné or television cameras, can be fitted into the project economically once the time of each activity is decided and the table of floats drawn up. Even if the network technique is not fully used the project leader will benefit vastly if he draws one for his own use. He will see more clearly what has to be done, by whom it must be done, when, and where, and it will enable him to delegate more efficiently.

Choosing audio-visual equipment

If the purchase of major items of equipment is well planned, costly mistakes can be avoided. The simple network in figure 63 includes the main activities and can easily be adapted to meet individual requirements.

The critical path has been indicated by a dotted line.

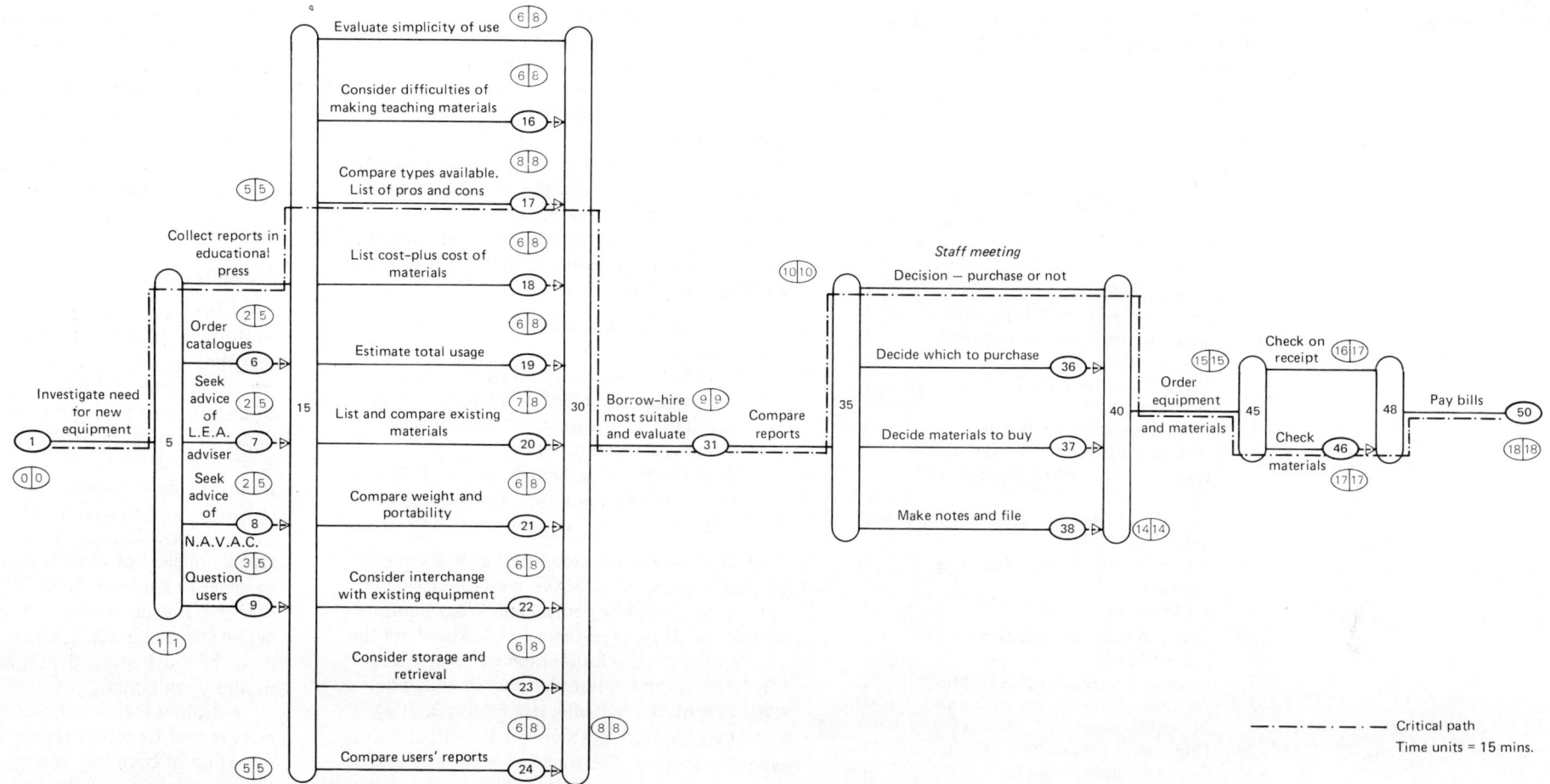

Fig. 63. *A network analysis of the choice of items of audio-visual equipment. In this network analysis a typical critical path has been identified.*

107

A NETWORK ANALYSIS OF THE CHOICE OF AUDIO-VISUAL EQUIPMENT

Activity number	Activity	Duration time (¼-hour units)
1–5	Investigate need for new equipment	1
5–15	Collect reports from educational press	4
5–6	Order catalogues	1
5–7	Seek advice from L.E.A. adviser	1
5–8	Seek advice from the National Audio Visual Aids Centre	1
5–9	Question those already using similar equipment	2
15–30	Evaluate simplicity of use	1
15–16	Consider difficulty of making teaching material to be used with equipment	1
15–17	Compare types available, list pros and cons	3
15–18	List cost, plus cost of teaching materials	1
15–19	Estimate total usage	1
15–20	List and compare existing teaching materials	2
15–21	Compare weight and portability	1
15–22	Consider interchange with existing equipment	1
15–23	Consider storage and retrieval	1
15–24	Compare users reports	1
30–31	Borrow/hire most suitable equipment and try out	1
31–35	Evaluate try-out	1
35–40	Staff meeting. Decide whether to buy	
35–36	Decide which to buy	4
35–37	Decide on materials to buy	
35–38	Make notes and file	
40–45	Order equipment and teaching materials	1
45–48	Check on receipt	1
45–46	Check materials on receipt	2
48–50	Pay bills and file	1

Closed-circuit television

During the past few years I have written and produced a number of closed circuit television (CCTV) programmes. I am keenly aware of some of the problems which can arise, of the time such productions can consume, and of the mistakes that can be made. It seemed a worthwhile exercise to apply the technique of network analysis.

Since CCTV production is a complex and costly operation it must be assumed that the following conditions are met:

1 That black and white moving pictures with sound are needed.
2 That the programme will be used many times, or in a number of schools, or that if it will not be used many times, or in many schools, it will not date.
3 That the teaching need cannot be met by a film, or slides and tape commentary, which might be cheaper, or by existing resources.

Once it has been decided that a CCTV programme is the most effective medium, existing constraints should be considered. What equipment is available? How experienced and skilled are the staff involved? How much time will everything take? This is the kind of planning in which network analysis is invaluable (figure 64, p. 109).

Not all the activities shown here will appear in every production. The time taken for any one of the activities shown will depend upon the skill and experience of those involved. An experienced and skilled team will take much less time, and conversely inexperienced, unskilled teams will take a good deal longer. The importance of early ordering of the materials needed should be brought home to a producer if the need can be seen at a glance.

Probably one of the most valuable attributes of this sort of planning is the benefit to the network designer. The drawing of a network involves the planner in a critical appraisal of a project right from the start, and then a logical building up of the plan to be used. In the critical appraisal the planner examines exactly what objectives are envisaged, and then sets out the way in which they are to be achieved. It may well be that the CCTV project will be set aside for a ciné film, which will achieve the same objective. If a CCTV programme is the best solution the planning by network analysis should ensure that the production is economical, in time, materials and manpower.

Teaching practice

Time spent by students in schools is valuable and so the utmost effort should be made to ensure that they derive the maximum benefit. Although conditions and schemes of work at colleges of education vary widely the network analysis in figure 65 can easily be adapted to suit particular needs. The aim is to show clearly to a student what is going to happen, to give a plan of what is to be done, and the time allocated to it. This last will of course depend upon the policy of an individual college but some indication should surely be given to students, especially on their first school practice. If each student knows exactly what he or she should be doing, and within what time limits he must work, then misunderstandings should be minimal.

In dealing with landladies who may or may not be prepared to accept students another year, confirmation of booking, and a letter of thanks after the school practice helps to smooth the way for next year's students. Similarly letters of thanks to members of staff at the school visited, and to the headmaster are very much appreciated — and are quite rare. Most head teachers and most class teachers go to considerable trouble to make sure that the time spent in school by students is used to the best advantage. The class is often disrupted by the introduction of a new approach and new methods, and a letter of thanks helps to smooth down any ruffled feathers and will help to make the effort worth while.

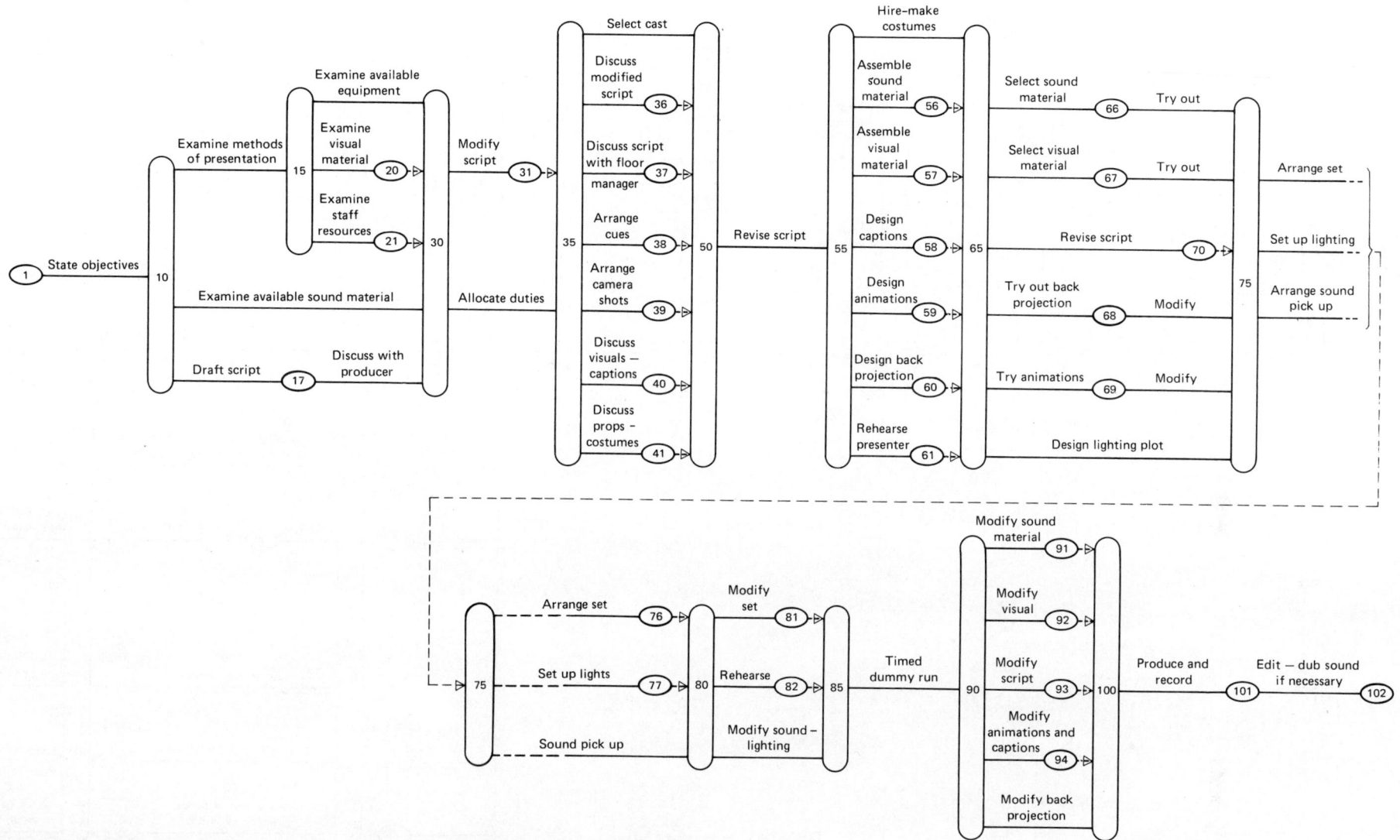

Fig. 64. *A network analysis of the production of a closed circuit television teaching programme.*

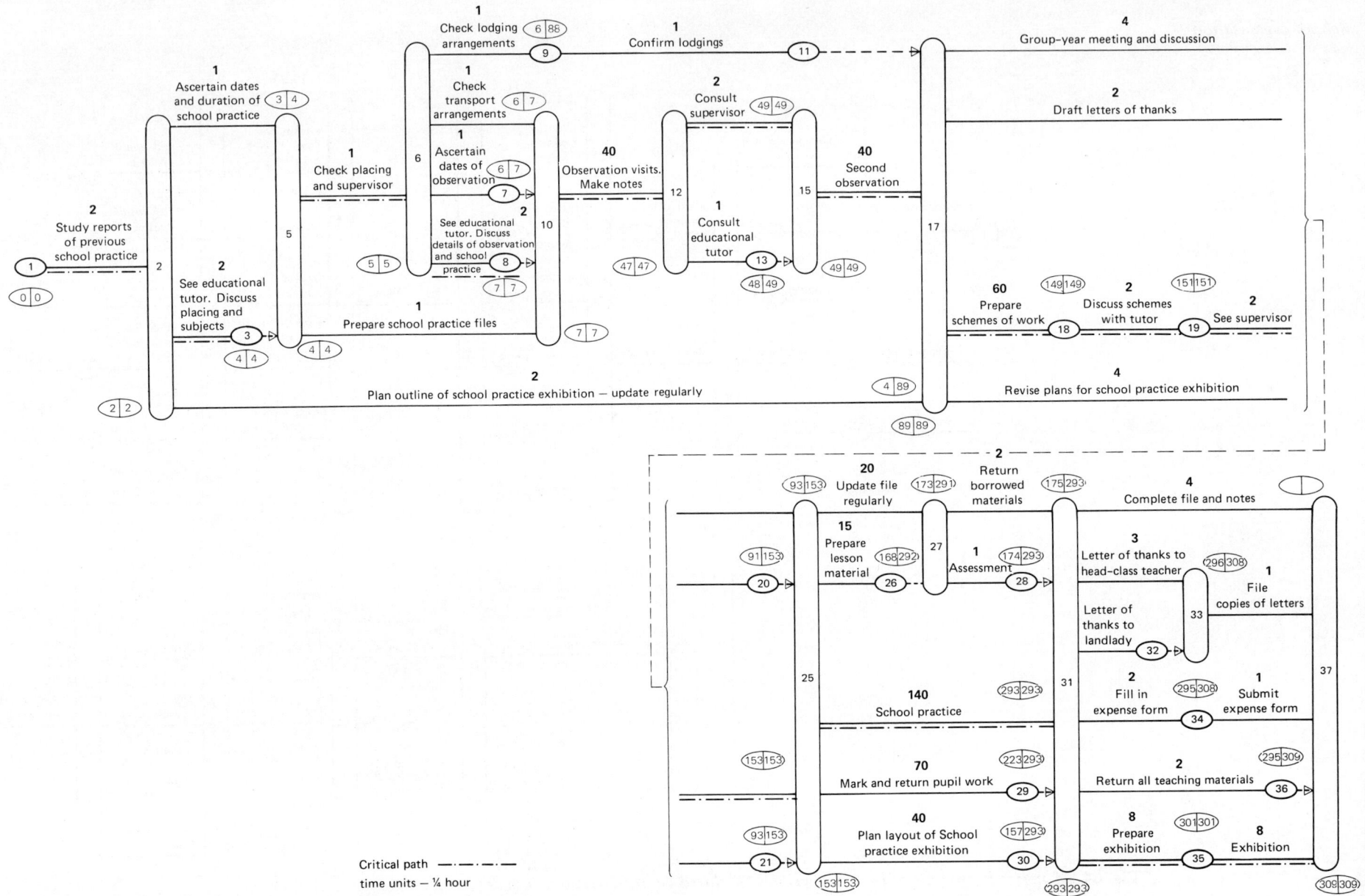

Fig. 65. *A network analysis of school practice during a course at a college of education. The probable critical path is shown as a dotted line.*

In their treatment of pupils the students should bear in mind that their work is important to them, and so it should be treated with respect. It should be displayed with tact and care, and marked thoughtfully. Any work needed for exhibition should be requested. Permission will almost always be freely given, but pupils will appreciate being asked if they will allow it to be taken away, and will be glad to get it back.

In dealing with staff at the resources centre, or library at college it helps if any materials needed are sought at an early date and not taken in a last minute rush. When materials have been borrowed they should be returned as soon as possible and in good order. The same remarks apply to equipment.

All the remarks about materials and equipment borrowed from the college apply, perhaps with even more force, to any materials or equipment borrowed from schools. Usually text books are freely lent to students to help them in their preparation and it is very irritating and time-consuming to try to recover them from students who may have gone on holiday. No text books in use in school are over-supplied and every one has to be accounted for. Replacing them is costly in time and money, if it is possible to replace them all. Hence the inclusion of activities about the return of materials and equipment.

The same care in returning items applies to copies of schemes of work, or curricula. These too are often freely lent, and it is hoped that care is taken to return them. Usually schemes of work consist of a number of duplicated sheets and it again is costly in time and materials to duplicate new ones.

There are reminders that time should be taken in visiting and consulting education tutors and supervisors. If this is built into any plan for the student it is not so likely to be overlooked or left until the last moment.

Another complementary network for the use of college staff who would be involved in a different series of activities could also be drawn.

ALLOCATION OF TIME – SCHOOL PRACTICE

Node numbers	Activity	Duration time (¼-hour units)
1–2	Study reports of previous school practice	2
2–5	Ascertain dates of school practice	1
2–3	See education tutor – placing and subjects	2
5–6	Check placing and tutor	1
5–10	Prepare school practice file	2
6–9	Check lodging arrangements	1
6–10	Check transport arrangements	1
6–7	Ascertain dates of observation	1
6–8	See education tutor, check details of observation and school practice	2
9–11	Confirm lodgings	1
10–12	Observation. Make full notes (2 days)	40
12–13	Consult supervisor about observation	2
12–15	Consult education tutor about observation	1
10–17	Collect/prepare visual materials for school practice	12
2–17	Plan rough outline of school practice exhibition	2
15–17	Second observation	40
17–25	Group or year meetings	4
17–20	Draft letters of thanks	2
17–18	Prepare schemes of work	60
17–21	Revise/modify plans for exhibition	4
18–19	Discuss schemes with education tutor and have them confirmed	2
19–25	See supervisor about scheme of work	2
25–27	Up-date file regularly	20
25–26	Prepare additional lesson materials	15
25–31	School practice (7½ weeks)	140
25–29	Mark and return all work done by pupils	70

Node numbers	Activity	Duration time (¼-hour units)
25–30	Plan layout of school practice exhibition	4
27–31	Return all borrowed materials	2
27–28	Assessment	1
31–33	Send letter of thanks to head teacher, and members of staff	3
31–32	Letter of thanks to landlady	1
31–34	Fill in expenses form	2
31–36	Return all teaching materials	2
31–35	Prepare exhibition	8
31–37	File and complete all notes	1
34–37	Submit expenses form	1
35–37	Exhibition	8

Solutions to problems in chapter 2

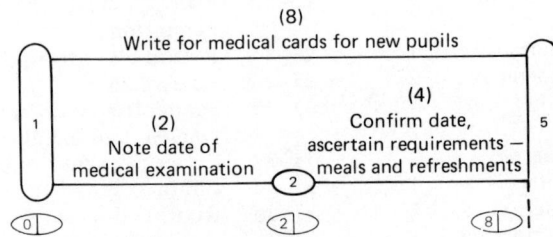

Fig. 66. *Possible solution to problem 1. Note that the writing for record cards can take place in parallel with the other activities. That the confirmation of the date must follow notification of the date. The time entered in the left-hand side of the time oval at the side of node number 5 will be 8 units as this is the larger of the possible times.*

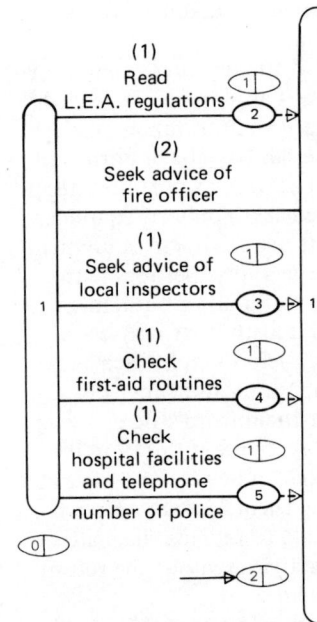

Fig. 67. *Possible solution to problem 2. Note that all these activities can take place in parallel. They could be carried out by different people at the same time. They need not be in the same positions from top to bottom so long as they are shown as parallel arrows.*

The time entered in the time oval at the foot of activity number 10 will be 2 units as this is the largest of the times in these activities.

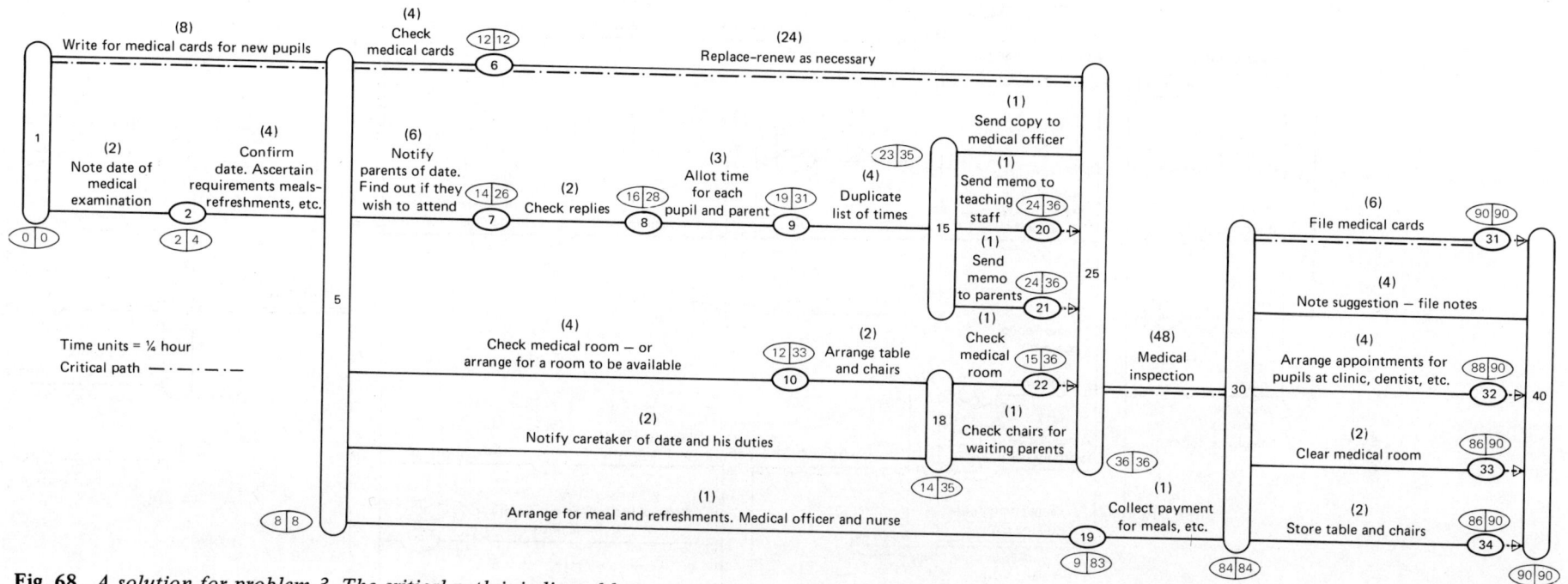

Fig. 68. *A solution for problem 3. The critical path is indicated by means of a dotted line. Your solution
may differ in detail but the critical path should be the same series of activities.*

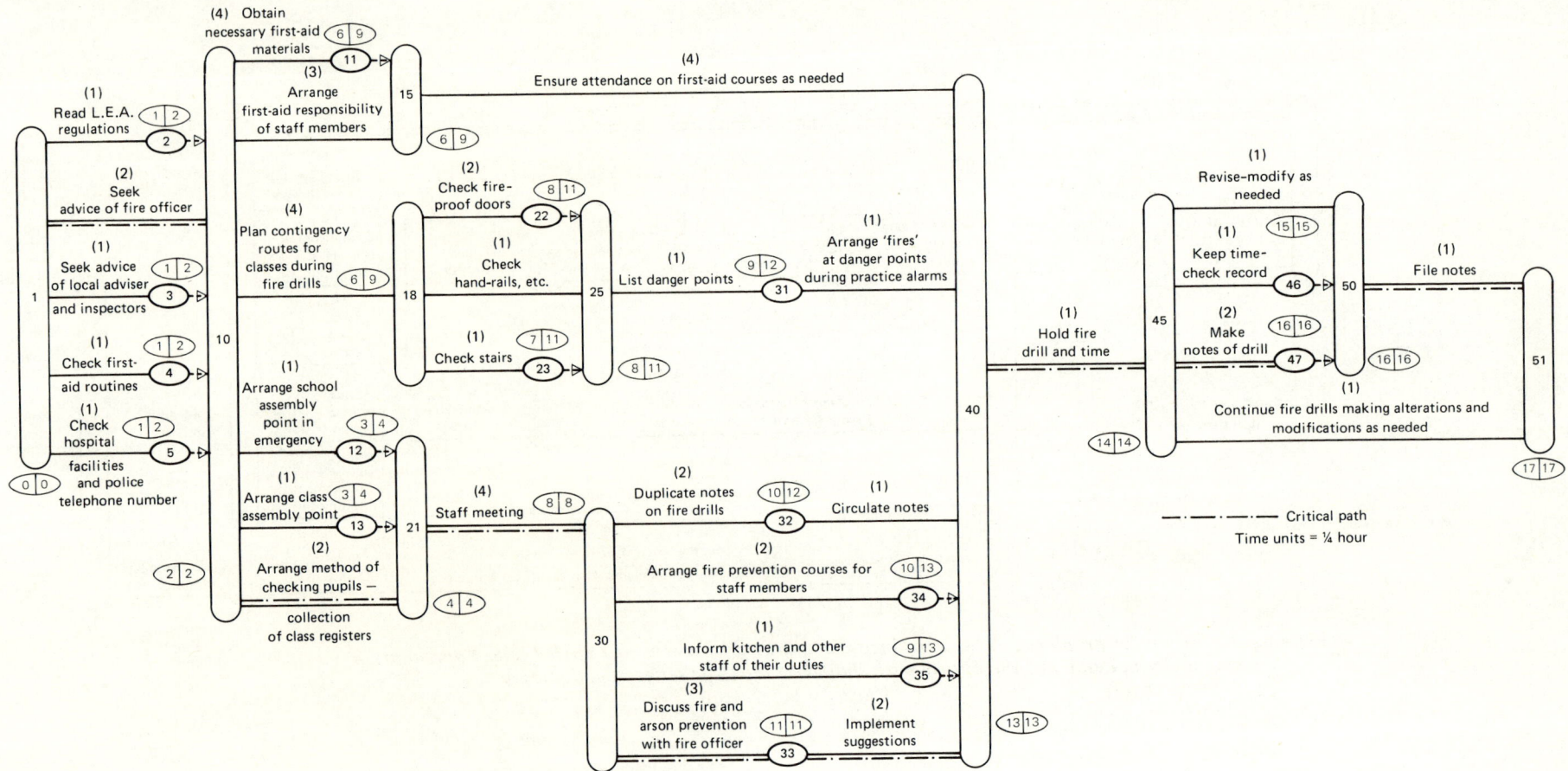

Fig. 69. *A solution to problem 4. Your solution may be different in detail but the critical path should consist of the same series of activities shown as a dotted line.*

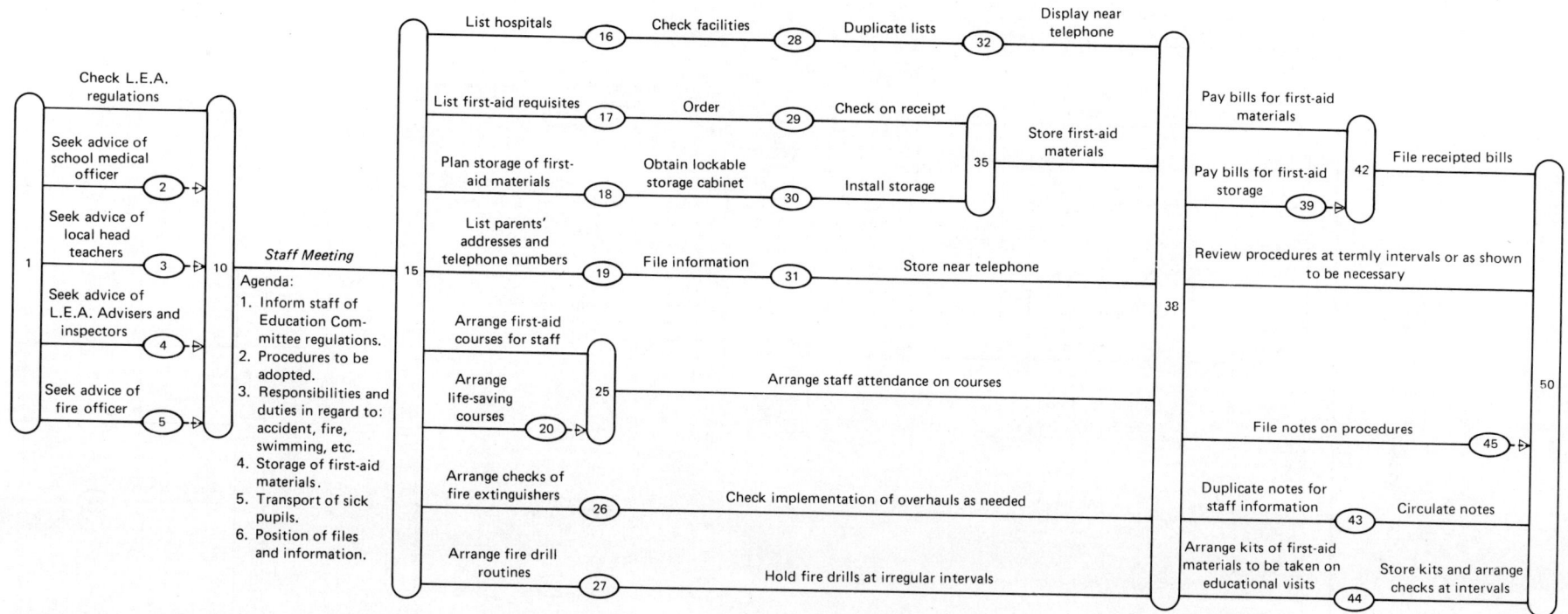

Fig. 70. *A network analysis of some of the activities which could appear on your solution to problem 5. Yours may differ considerably for regulations will vary in each education authority.*

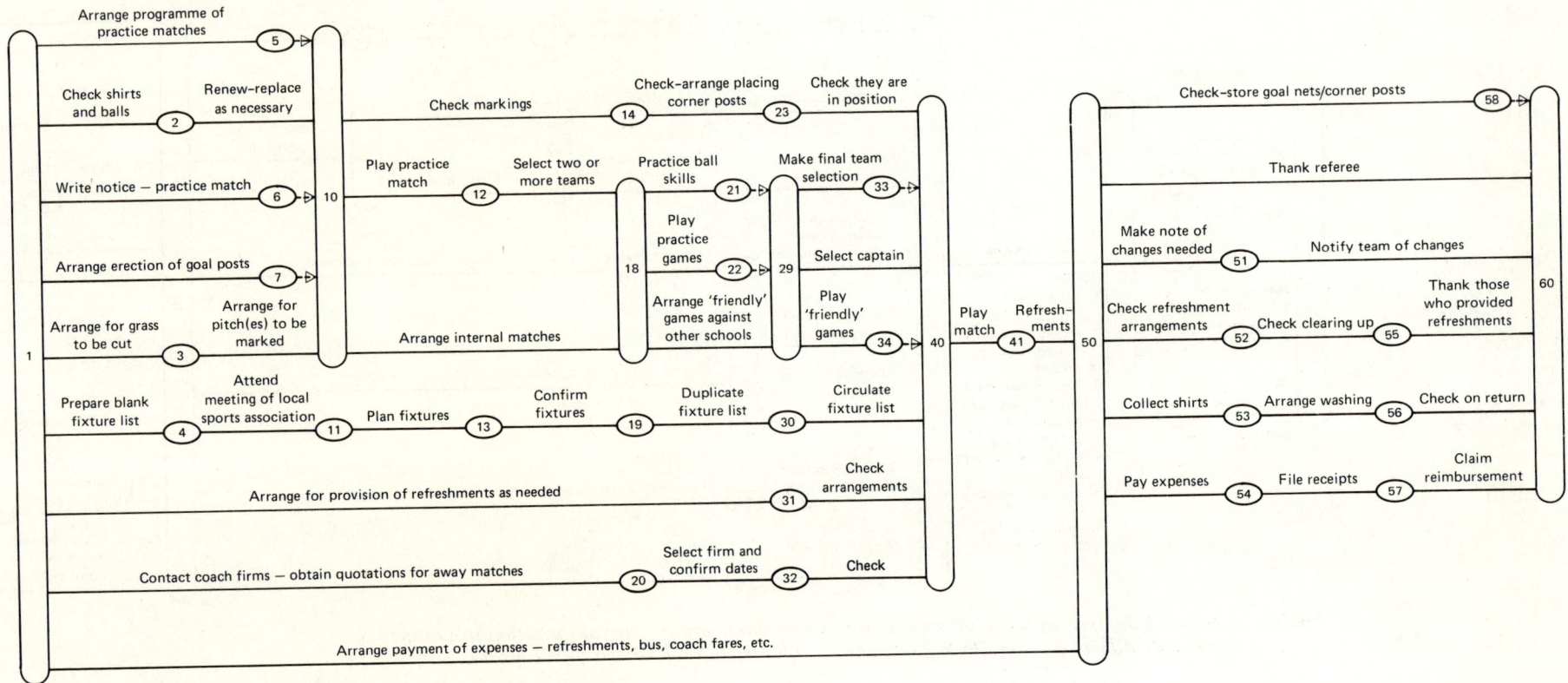

Fig. 71. *A solution to problem 6. Your solution may be different as conditions vary widely from school to school.*

Confirm date with kitchen staff — 20 — Arrange use of kitchen

Confirm date with P.T.A. refreshment committee — 9 — Arrange supply of food and drink — 21 — Arrange supply of crockery and cutlery — 31 — Check arrangements

Discuss meeting with school staff — 6 — Inform school caretaker of meeting — 22 — Give details of chairs and tables required — 32 — Check arrangements

Discuss meeting with Hon. Secretary of the P.T.A. — 12

Discuss meeting with Chairman of the P.T.A. — 7 — Arrange evening's programme — 23 — Duplicate notices of meeting — 33 — Circulate

Give details about speaker to chairman for his introduction

Read minutes of previous meetings — 1 — 5

Confirm meeting with speaker — 13 — Check his requirement of audio-visual equipment — 24 — Arrange audio-visual equipment as needed — 34 — Check audio-visual equipment — 40

Arrange for vote of thanks for speaker (bouquet for his wife)

Discuss meeting with Hon. Treasurer — 14 — Arrange for payment of speaker's expenses — also refreshments

Contact Chairman — arrange for presentation of his Report — Arrange method of election of officers — 25 — Collect nominations — 35 — Arrange for supply of paper and pencils

Arrange for votes of thanks: Chairman, Hon. Secretary and Hon. Treasurer — 18 — 8

Contact Hon. Treasurer — arrange for presentation of his Report

Contact Hon. Secretary — arrange for presentation of his Report — 19 — 30 — Duplicate as needed — 36 — Arrange distribution

Draft letter of thanks to speaker

P.T.A. meeting — 45
1. Introduction — speaker.
2. Speaker.
3. Questions.
4. Vote of thanks.
5. Refreshments.
6. Notices.
7. A.O.B.

A.G.M.
1. Chairman's Report.
2. Vote of thanks.
3. Hon. Secretary's Report.
4. Vote of thanks.
5. Hon. Treasurer's Report.
6. Vote of thanks.
7. Election of Officers.
8. Plan year's programme.
9. Names for committees.
10. A.O.B.

Arrange–circulate minutes of meeting and notice of next — 46 — Duplicate — 50 — Circulate

Thank all officers of P.T.A. — 55

Thank refreshment committee — 51 — Check kitchen

Send confirmation of appointments — 47 — Check replies — 52 — File replies

Check speaker has received his expenses

Check payment of expenses– refreshments — 48 — Collect paid accounts — 53 — File

Send letter of thanks to speaker — 49 — Check reply — 54 — File — 60

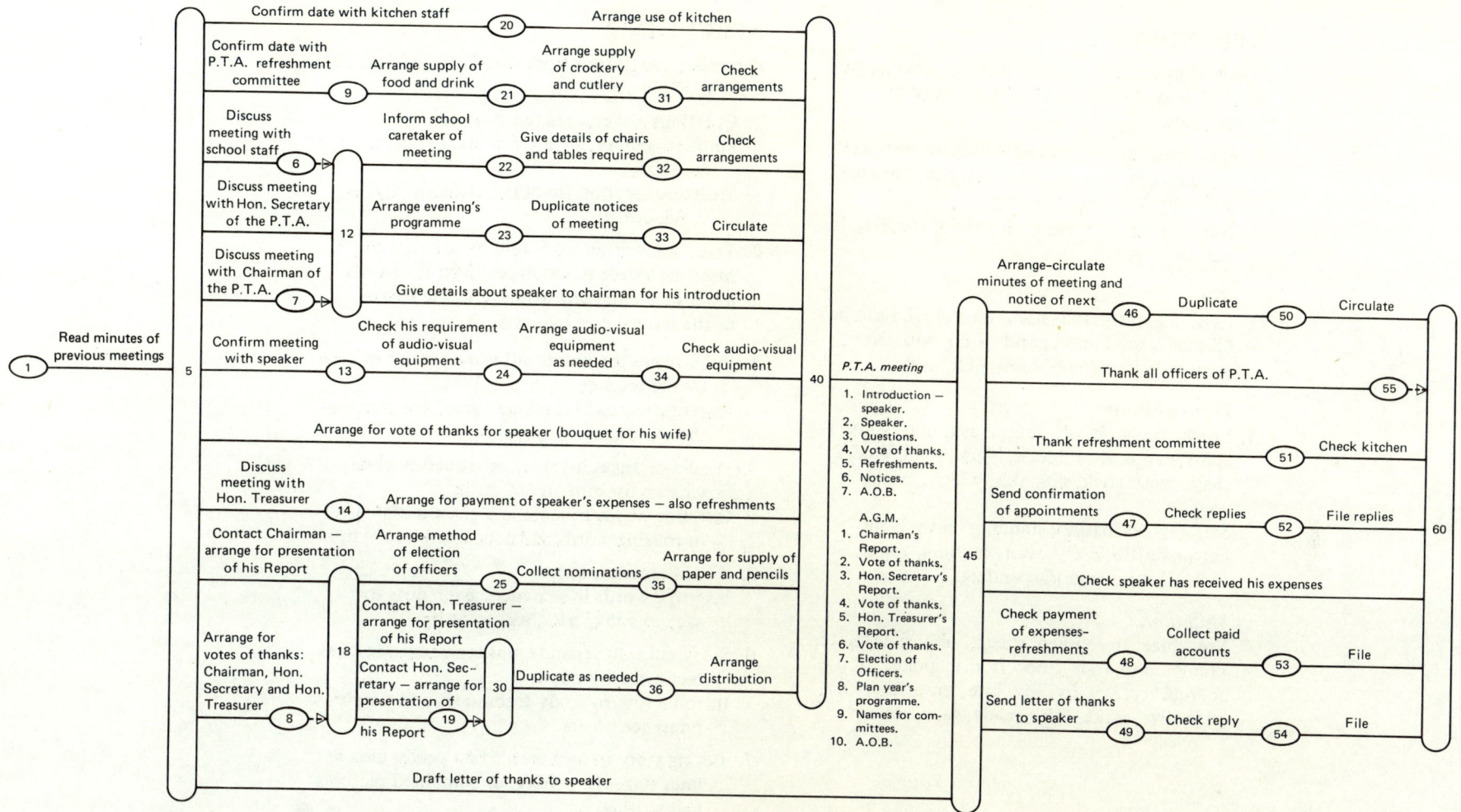

Fig. 72. *A network analysis of typical solution to problem 7. Once again the details may vary but overall content should be similar.*

APPENDIX 1

Possible progression for phonic elements (see pp. 68, 70). Nos. 285, 286, 287 on network of curriculum.

1. Man, can, cat, ran, rat, had, bed, let, hen, get, beg, pig, did, him, sit, win, hot, top, box, not, nod, sun, but, bug, run, rug.
2. Hut, hot, hit, hat, ham, him, hem, beg, bag, big, bug, bit, bat.

Blends
3. Pass, flag, fast, sand, stem, went, bell, best, ball, bill, pets, spin, milk, pond, wind, will, from, stop, doll, just, jump, hunt, dull, dust.

Double blends
4. Stuck, chunk, blush, shut, crash, that, strap, spots, chops, shot, block, black, shall, shed, shall, chest, stick, sing, this, which, chick, chop.
5. Sally, under, basket, standing, until, little, battle, bottle, Betty, river, running, apple, better, kitty, cannot, spending.

Mutations
6. See, three, tree, green, sheep, tube, rude, flute, plume, note, rode, hope, home, spoke, go, no, so, my, try, sky, by, spy, bite, time, like, kite, why, dive, make, cake, made, gave, etc.

APPENDIX 2

Progression suggested from nos. 281, 282 and 283 (see pp. 68, 70).

1. Questions about a related story.
    Prefixes: sub, re, ab, ad, pro, ex, be, com, dis, de, en.
    Suffixes: ure, ous, ion, tion, ation, ly, ity, er, y, ful, ent, al.

2. Tracing a written story told by another pupil.
    Inserting words in sentences given the initial letter of the missing words, and the shapes of the letters, e.g. l☐☐ = lip.

3. Pupil traces the words of his own story written by the teacher.
    Inserting words in sentences given the shape of the words, e.g. ☐☐☐ = lip.

4. Pupil rearranges his story in sequence when written for him on cut cards.
    Inserting words in sentences given initial letters of missing words and dots, e.g. l . . . = lip.

5. 'Flash' sentence strips.
    Inserting words in sentences given dots for each missing letter, e.g. . . . = lip.

6. Story cut into separate words arranged in correct order by pupil.
    Introduction to study techniques and simple reference books.

7. Telling story in own words first orally then at later stages in writing, or combined pictures and writing.
    'Surveying' techniques. Book titles, author, chapter headings, illustrations, summary in dust cover, etc.

8. Making notes about books read, answering questions about books read.

9. Study techniques. S.Q.3R (survey, question, read, recite aloud, revise).

Bibliography

The operational research technique of network analysis

Battersby, A. *Network Analysis,* Macmillan, 1970
Lockyer, K. G. *An Introduction to Critical Path Analysis*, Pitman, 1969

Network analysis used in education

Dowling, T. J. 'Network Analysis as an Aid to Course Planning', *The Vocational Aspect of Education*, Vol. XXVI, No. 65
Platts, C. V. & Wyant, T. G. 'Network Analysis and the possibility of its use in Education', *Educational Review*, 21 Feb. 1969
Stewart, J. D. *Local Government Chronicle*, Feb. 1969
Vaughan, B. W. 'Network Analysis in Primary School Mathematics', *Science Teacher*, Dec. 1971
  'The Application of the Operational Research Technique of Network Analysis to Primary Mathematics', in *Aspects of Educational Technology*, VI, Pitman, 1972
  'The Application of a Modified Operational Research Technique in the design of curricula. Systematic Curriculum Design (SCUD) Technique', in *Aspects of Educational Technology*, XI, Kogan Page, 1978
Werts, C. E. 'The Study of College Environment using Path Analysis', *Research Reports*, Vol. 3, No. 4, 1967
Wood, A. & Wyant, T. G. 'Using Network Analysis on a learning Sequence', *Industrial and Commercial Training*, Dec. 1970
Wyant, T. G. 'Critical Path Analysis of a Course', *Educational Review,* Vol. 2, No. 2, Feb. 1967
  'Course Analysis', in *Aspects of Educational Technology*, VII, Pitman, 1973

Index